# Linking Genetic Resources and Geography: Emerging Strategies for Conserving and Using Crop Biodiversity

# Linking Genetic Resources and Geography: Emerging Strategies for Conserving and Using Crop Biodiversity

Proceedings of a symposium cosponsored by Division C-8 of the Crop Science Society of America, Division A-6 of the American Society of Agronomy, and ERDAS, Inc., in Anaheim, CA, 29 Oct. 1997.

*Editors*
Stephanie L. Greene and Luigi Guarino

*Organizing Committee*
Stephanie L. Greene and Thomas C. Hart

*Editorial Committee*
Stephanie L. Greene and Luigi Guarino

*Editor-in-Chief ASA*
Jerry Hatfield

*Editor-in-Chief CSSA*
Jeff Volenec

*Managing Editor*
J.M. Bartels


CSSA Special Publication Number 27


American Society of Agronomy, Inc.
Crop Science Society of America, Inc.
Madison, Wisconsin, USA

1999

The cover illustrates a classic example of the spatial distribution of genetic diversity, which is the geographical distribution of spikelet shape classes in cultivated sorghum: Durra, adapted to drier Arabian habitats; Kafir, found mainly south of the equator in Africa; and Guinea, adapted to the wetter habitats of western Africa (see Harlan & de Wet in Crop Science, Volume 12, p. 172–176).

American Society of Agronomy, Inc.
Crop Science Society of America, Inc.
677 South Segoe Road, Madison, WI 53711 USA

Library of Congress Registration Number 99-072729

Printed in the United States of America

# CONTENTS

Foreword . . . . . . . . . . . . . . . . . . . . . . . . . . . . . . . . . . . . . . . . . . . . . . . . . . . . . . . . . .

Preface . . . . . . . . . . . . . . . . . . . . . . . . . . . . . . . . . . . . . . . . . . . . . . . . . . . . . . . . . . . .

Contributors . . . . . . . . . . . . . . . . . . . . . . . . . . . . . . . . . . . . . . . . . . . . . . . . . . . . . . .

Acknowledgment . . . . . . . . . . . . . . . . . . . . . . . . . . . . . . . . . . . . . . . . . . . . . . . . . . .

Conversion Factors for SI and Non-SI Units . . . . . . . . . . . . . . . . . . . . . . . . .

1    Analysis of Georeferenced Data and the Conservation
        and Use of Plant Genetic Resources
                Luigi Guarino, Nigel Maxted, and Mark C. Sawkins . . . . . . . . .

2    Implementing Geographic Analysis in Germplasm Conservation
                Stephanie L. Greene and Thomas C. Hart . . . . . . . . . . . . . . . .

3    Exploring the Relationship of Plant Genotype and
        Phenotype to Ecogeography
                Jeffrey J. Steiner . . . . . . . . . . . . . . . . . . . . . . . . . . . . . . . . . . .

4    Scale Considerations in Mapping for Germplasm
        Acquisition and the Assessment of Ex Situ Collections
                Thomas C. Hart . . . . . . . . . . . . . . . . . . . . . . . . . . . . . . . . . . . . .

5    Matching Germplasm to Geography :Environmental
        Analysis for Plant Introduction
                Trevor H. Booth . . . . . . . . . . . . . . . . . . . . . . . . . . . . . . . . . . . .

6    Germplasm Collecting Using Modern Geographic
        Information Technologies: Directions Explored
        by the N.I. Vavilov Institute of Plant Industry
                Alexandr Afonin and Stephanie L. Greene . . . . . . . . . . . . . . .

7    Predicting Species Distributions Using Environmental
        Data: Case Studies Using *Stylosanthes* Sw.
                Mark C. Sawkins, Peter G. Jones, Nigel Maxted,
                Roger Smith, and Luigi Guarino . . . . . . . . . . . . . . . . . . . . . . . .

8    Institutional Adoption of Spatial Analytical Procedures:
        Where Is the Bottleneck?
                John D. Corbett and Paul Dyke . . . . . . . . . . . . . . . . . . . . . . . .

# FOREWORD

The genetic diversity of plants is affected by the climate and landscape of the geographic location. With the advent of the computer-based geographic information system (GIS) as a tool for identifying spatial information, it is now possible to apply the GIS technology to map the distribution of plant genetic resources in relation to the climatic and geographic characteristics of the locale. The GIS technology provides an efficient means to link genetic resources to geographic locations. The ability to link diverse genetic materials to landscape characteristics will greatly enhance efforts both to explore biodiversity as a strategy for crop adaptation and to understand how climate and landscape changes would affect the adaptability of crops to specific geographic regions. More efficient genetic conservation strategies will be more beneficial to future scientists using genetic resources for plant improvement.

We appreciate the foresight and effort of Drs. Stephanie Greene and Luigi Guarino in organizing the symposium, from which this special publication derives. The symposium provided an opportunity to examine the feasibility of using the GIS tools for developing strategies for conserving and using crop biodiversity effectively. This publication will serve as a guide for those working with genetic resources in collecting and managing environmental datasets with GIS tools. It will be an important addition to the libraries of genetic scientists as well as to teaching, research, and industrial institutions.

H.H. CHENG, *president*
*American Society of Agronomy*

L.E. MOSER, *president*
*Crop Science Society of America*

# PREFACE

Geography, through climate and landscape, influences natural selection and gene flow, key processes determining the extent, and shaping the pattern, of genetic diversity within taxa and gene pools. Recognizing this relationship, plant genetic resource workers always have used the geographic information available to them to guide the exploration, collecting and use of genetic diversity. This information has generally taken the form of paper maps and narrative descriptions. However, with recent technical innovations in the field of geography and cartography, more information on the environment is available in a digital format. Geographic Information System (GIS) technology provides tools for the effective and efficient manipulation and analysis of such spatially referenced digital information.

The wealth of environmental information and availability of user-friendly GIS tools empowers plant genetic resource workers in their exploitation of the linkage between genetic diversity and geography. In anticipation of the profound impact these emerging strategies could have, a symposium was organized on the occasion of the 89th ASA-CSSA-SSSA Annual Meeting in Anaheim, California, on 26 to 30 Oct. 1997. The objectives of the symposium were to: (i) review the application of spatial analysis to genetic resource conservation and use issues, and (ii) discuss the successful implementation of geographic analysis in individual projects as well as the institutional adoption of GIS practice.

This publication is the proceedings of the Anaheim symposium. It represents the first attempt to provide a framework for applying GIS applications to genetic resource issues, supported by a wide range of examples. It also discusses some of the specific barriers to the transfer and adoption of this technology by institutions concerned with the conservation and use of plant genetic resources. We hope that it will provide an impetus to national plant genetic resources programs the world over to investigate the use of GIS tools and digital environmental datasets to develop conservation strategies that are effective, scientifically sound and efficient of scarce resources.

STEPHANIE L. GREENE, *Editor*
*Washington State University*
*Prosser, Washington*

LUIGI GUARINO, *Editor*
*International Plant Genetic Resources Institute*
*Cali, Colombia*

# ACKNOWLEDGMENT

We are grateful to the authors for their efforts and enthusiasm and hard work to complete this project. We would like to extent our thanks to T.C. Hart, of Spatial Associates, Inc., for helping to organize the symposium and providing so freely of his time. We also appreciate the time spent by the reviewers of the various papers. Their thoughtful comments have been helpful in ensuring this publication will make a valuable contribution to the conservation and use of plant genetic resources. We would like to acknowledge the C-8 and A-6 Divisions of the CSSA and ASA Societies and ERDAS, Inc., a GIS software company, for the support that allowed us to bring together the speakers for the symposium. Finally, we would like to thank the Madison Headquarters staff for producing this publication.

# CONTRIBUTORS

**Alexandr Afonin**

N.I. Vavilov Institute of Plant Industry, Bolshaya Morskaya 42, St. Petersburg, Russia 190000

**Trevor H. Booth**

CSIRO Forestry and Forest Products, PO Box E4008, Kingston, Canberra, ACT 2604, Australia

**John D. Corbett**

Blackland Research Center, Texas A & M University, 808 E. Blackland Road, Temple, TX 76502

**Paul Dyke**

Blackland Research Center, Texas A & M University, 808 E. Blackland Road, Temple, TX 76502

**Stephanie L. Greene**

Department of Crop and Soil Science, Washington State University, 24106 N. Bunn Road, Prosser, WA 99350

**Luigi Guarino**

Regional Office for the Americas, International Plant Genetic Resources Institute, c/o CIAT, Apartado Aereo 6713, Cali, Colombia

**Thomas C. Hart**

Spatial Data Associates, 336 Pennsylvania Ave., Trumensburg, NY, 14885

**Peter G. Jones**

Centro Internacional de Agricultura Tropicale (CIAT), Apartado Aereo 6713, Cali, Colombia

**Nigel Maxted**

School of Biological Sciences, University of Birmingham, Edgbaston, Birmingham, B15 2TT, United Kingdom

**Mark C. Sawkins**

School of Biological Sciences, University of Birmingham, Edgbaston, Birmingham, B15 2TT, United Kingdom; and The Royal Botanic Gardens, Kew, Richmond, TW9 3D5, United Kingdom

**Roger Smith**

RBG Kew Seed Bank, Wakehurst Place, Ardingly, Haywards Heath, West Sussex, RH17 6TN, United Kingdom

**Jeffrey J. Steiner**

National Forage Seed Production Research Center, USDA Agricultural Research Service, 3450 SW Campus Way, Corvallis, OR, USA 97331

# Conversion Factors for SI and non-SI Units

# Conversion Factors for SI and non-SI Units

| To convert Column 1 into Column 2, multiply by | Column 1 SI Unit | Column 2 non-SI Units | To convert Column 2 into Column 1, multiply by |
|---|---|---|---|
| **Length** | | | |
| 0.621 | kilometer, km ($10^3$ m) | mile, mi | 1.609 |
| 1.094 | meter, m | yard, yd | 0.914 |
| 3.28 | meter, m | foot, ft | 0.304 |
| 1.0 | micrometer, $\mu$m ($10^{-6}$ m) | micron, $\mu$ | 1.0 |
| $3.94 \times 10^{-2}$ | millimeter, mm ($10^{-3}$ m) | inch, in | 25.4 |
| 10 | nanometer, nm ($10^{-9}$ m) | Angstrom, Å | 0.1 |
| **Area** | | | |
| 2.47 | hectare, ha | acre | 0.405 |
| 247 | square kilometer, km² $(10^3$ m$)^2$ | acre | $4.05 \times 10^{-3}$ |
| 0.386 | square kilometer, km² $(10^3$ m$)^2$ | square mile, mi² | 2.590 |
| $2.47 \times 10^{-4}$ | square meter, m² | acre | $4.05 \times 10^3$ |
| 10.76 | square meter, m² | square foot, ft² | $9.29 \times 10^{-2}$ |
| $1.55 \times 10^{-3}$ | square millimeter, mm² $(10^{-3}$ m$)^2$ | square inch, in² | 645 |
| **Volume** | | | |
| $9.73 \times 10^{-3}$ | cubic meter, m³ | acre-inch | 102.8 |
| 35.3 | cubic meter, m³ | cubic foot, ft³ | $2.83 \times 10^{-2}$ |
| $6.10 \times 10^4$ | cubic meter, m³ | cubic inch, in³ | $1.64 \times 10^{-5}$ |
| $2.84 \times 10^{-2}$ | liter, L ($10^{-3}$ m³) | bushel, bu | 35.24 |
| 1.057 | liter, L ($10^{-3}$ m³) | quart (liquid), qt | 0.946 |
| $3.53 \times 10^{-2}$ | liter, L ($10^{-3}$ m³) | cubic foot, ft³ | 28.3 |
| 0.265 | liter, L ($10^{-3}$ m³) | gallon | 3.78 |
| 33.78 | liter, L ($10^{-3}$ m³) | ounce (fluid), oz | $2.96 \times 10^{-2}$ |
| 2.11 | liter, L ($10^{-3}$ m³) | pint (fluid), pt | 0.473 |

## Mass

| To convert Column 1 into Column 2, multiply by | Column 1 SI Unit | Column 2 non-SI Unit | To convert Column 2 into Column 1, multiply by |
|---|---|---|---|
| $2.20 \times 10^{-3}$ | gram, g ($10^{-3}$ kg) | pound, lb | 454 |
| $3.52 \times 10^{-2}$ | gram, g ($10^{-3}$ kg) | ounce (avdp), oz | 28.4 |
| 2.205 | kilogram, kg | pound, lb | 0.454 |
| 0.01 | kilogram, kg | quintal (metric), q | 100 |
| $1.10 \times 10^{-3}$ | kilogram, kg | ton (2000 lb), ton | 907 |
| 1.102 | megagram, Mg (tonne) | ton (U.S.), ton | 0.907 |
| 1.102 | tonne, t | ton (U.S.), ton | 0.907 |

## Yield and Rate

| To convert Column 1 into Column 2, multiply by | Column 1 SI Unit | Column 2 non-SI Unit | To convert Column 2 into Column 1, multiply by |
|---|---|---|---|
| 0.893 | kilogram per hectare, kg ha$^{-1}$ | pound per acre, lb acre$^{-1}$ | 1.12 |
| $7.77 \times 10^{-2}$ | kilogram per cubic meter, kg m$^{-3}$ | pound per bushel, lb bu$^{-1}$ | 12.87 |
| $1.49 \times 10^{-2}$ | kilogram per hectare, kg ha$^{-1}$ | bushel per acre, 60 lb | 67.19 |
| $1.59 \times 10^{-2}$ | kilogram per hectare, kg ha$^{-1}$ | bushel per acre, 56 lb | 62.71 |
| $1.86 \times 10^{-2}$ | kilogram per hectare, kg ha$^{-1}$ | bushel per acre, 48 lb | 53.75 |
| 0.107 | liter per hectare, L ha$^{-1}$ | gallon per acre | 9.35 |
| 893 | tonne per hectare, t ha$^{-1}$ | pound per acre, lb acre$^{-1}$ | $1.12 \times 10^{-3}$ |
| 893 | megagram per hectare, Mg ha$^{-1}$ | pound per acre, lb acre$^{-1}$ | $1.12 \times 10^{-3}$ |
| 0.446 | megagram per hectare, Mg ha$^{-1}$ | ton (2000 lb) per acre, ton acre$^{-1}$ | 2.24 |
| 2.24 | meter per second, m s$^{-1}$ | mile per hour | 0.447 |

## Specific Surface

| To convert Column 1 into Column 2, multiply by | Column 1 SI Unit | Column 2 non-SI Unit | To convert Column 2 into Column 1, multiply by |
|---|---|---|---|
| 10 | square meter per kilogram, m$^2$ kg$^{-1}$ | square centimeter per gram, cm$^2$ g$^{-1}$ | 0.1 |
| 1000 | square meter per kilogram, m$^2$ kg$^{-1}$ | square millimeter per gram, mm$^2$ g$^{-1}$ | 0.001 |

## Pressure

| To convert Column 1 into Column 2, multiply by | Column 1 SI Unit | Column 2 non-SI Unit | To convert Column 2 into Column 1, multiply by |
|---|---|---|---|
| 9.90 | megapascal, MPa ($10^6$ Pa) | atmosphere | 0.101 |
| 10 | megapascal, MPa ($10^6$ Pa) | bar | 0.1 |
| 1.00 | megagram, per cubic meter, Mg m$^{-3}$ | gram per cubic centimeter, g cm$^{-3}$ | 1.00 |
| $2.09 \times 10^{-2}$ | pascal, Pa | pound per square foot, lb ft$^{-2}$ | 47.9 |
| $1.45 \times 10^{-4}$ | pascal, Pa | pound per square inch, lb in$^{-2}$ | $6.90 \times 10^3$ |

(continued on next page)

# Conversion Factors for SI and non-SI Units

| To convert Column 1 into Column 2, multiply by | Column 1 SI Unit | Column 2 non-SI Units | To convert Column 2 into Column 1, multiply by |
|---|---|---|---|
| | | **Temperature** | |
| 1.00 (K − 273) | kelvin, K | Celsius, °C | 1.00 (°C + 273) |
| (9/5 °C) + 32 | Celsius, °C | Fahrenheit, °F | 5/9 (°F − 32) |
| | | **Energy, Work, Quantity of Heat** | |
| $9.52 \times 10^{-4}$ | joule, J | British thermal unit, Btu | $1.05 \times 10^{3}$ |
| 0.239 | joule, J | calorie, cal | 4.19 |
| $10^{7}$ | joule, J | erg | $10^{-7}$ |
| 0.735 | joule, J | foot-pound | 1.36 |
| $2.387 \times 10^{-5}$ | joule per square meter, J m$^{-2}$ | calorie per square centimeter (langley) | $4.19 \times 10^{4}$ |
| $10^{5}$ | newton, N | dyne | $10^{-5}$ |
| $1.43 \times 10^{-3}$ | watt per square meter, W m$^{-2}$ | calorie per square centimeter minute (irradiance), cal cm$^{-2}$ min$^{-1}$ | 698 |
| | | **Transpiration and Photosynthesis** | |
| $3.60 \times 10^{-2}$ | milligram per square meter second, mg m$^{-2}$ s$^{-1}$ | gram per square decimeter hour, g dm$^{-2}$ h$^{-1}$ | 27.8 |
| $5.56 \times 10^{-3}$ | milligram (H$_2$O) per square meter second, mg m$^{-2}$ s$^{-1}$ | micromole (H$_2$O) per square centimeter second, μmol cm$^{-2}$ s$^{-1}$ | 180 |
| $10^{-4}$ | milligram per square meter second, mg m$^{-2}$ s$^{-1}$ | milligram per square centimeter second, mg cm$^{-2}$ s$^{-1}$ | $10^{4}$ |
| 35.97 | milligram per square meter second, mg m$^{-2}$ s$^{-1}$ | milligram per square decimeter hour, mg dm$^{-2}$ h$^{-1}$ | $2.78 \times 10^{-2}$ |
| | | **Plane Angle** | |
| 57.3 | radian, rad | degrees (angle), ° | $1.75 \times 10^{-2}$ |

## Electrical Conductivity, Electricity, and Magnetism

| To convert Column 2 into Column 1, multiply by | Column 1 SI Unit | Column 2 non-SI Unit | To convert Column 1 into Column 2, multiply by |
|---|---|---|---|
| $10$ | siemen per meter, S m$^{-1}$ | millimho per centimeter, mmho cm$^{-1}$ | $0.1$ |
| $10^4$ | tesla, T | gauss, G | $10^{-4}$ |

## Water Measurement

| To convert Column 2 into Column 1, multiply by | Column 1 SI Unit | Column 2 non-SI Unit | To convert Column 1 into Column 2, multiply by |
|---|---|---|---|
| $9.73 \times 10^{-3}$ | cubic meter, m$^3$ | acre-inch, acre-in | $102.8$ |
| $9.81 \times 10^{-3}$ | cubic meter per hour, m$^3$ h$^{-1}$ | cubic foot per second, ft$^3$ s$^{-1}$ | $101.9$ |
| $4.40$ | cubic meter per hour, m$^3$ h$^{-1}$ | U.S. gallon per minute, gal min$^{-1}$ | $0.227$ |
| $8.11$ | hectare meter, ha m | acre-foot, acre-ft | $0.123$ |
| $97.28$ | hectare meter, ha m | acre-inch, acre-in | $1.03 \times 10^{-2}$ |
| $8.1 \times 10^{-2}$ | hectare centimeter, ha cm | acre-foot, acre-ft | $12.33$ |

## Concentrations

| To convert Column 2 into Column 1, multiply by | Column 1 SI Unit | Column 2 non-SI Unit | To convert Column 1 into Column 2, multiply by |
|---|---|---|---|
| $1$ | centimole per kilogram, cmol kg$^{-1}$ | milliequivalent per 100 grams, meq 100 g$^{-1}$ | $1$ |
| $0.1$ | gram per kilogram, g kg$^{-1}$ | percent, % | $10$ |
| $1$ | milligram per kilogram, mg kg$^{-1}$ | parts per million, ppm | $1$ |

## Radioactivity

| To convert Column 2 into Column 1, multiply by | Column 1 SI Unit | Column 2 non-SI Unit | To convert Column 1 into Column 2, multiply by |
|---|---|---|---|
| $2.7 \times 10^{-11}$ | becquerel, Bq | curie, Ci | $3.7 \times 10^{10}$ |
| $2.7 \times 10^{-2}$ | becquerel per kilogram, Bq kg$^{-1}$ | picocurie per gram, pCi g$^{-1}$ | $37$ |
| $100$ | gray, Gy (absorbed dose) | rad, rd | $0.01$ |
| $100$ | sievert, Sv (equivalent dose) | rem (roentgen equivalent man) | $0.01$ |

## Plant Nutrient Conversion

| To convert Column 2 into Column 1, multiply by | Elemental | Oxide | To convert Column 1 into Column 2, multiply by |
|---|---|---|---|
| $2.29$ | P | P$_2$O$_5$ | $0.437$ |
| $1.20$ | K | K$_2$O | $0.830$ |
| $1.39$ | Ca | CaO | $0.715$ |
| $1.66$ | Mg | MgO | $0.602$ |

# 1

# Analysis of Georeferenced Data and the Conservation and Use of Plant Genetic Resources

**L. Guarino**

*Regional Office for the Americas*
*International Plant Genetic Resources Institute*
*Cali, Colombia*

**N. Maxted and M. Sawkins**

*University of Birmingham*
*Birmingham, United Kingdom*

Countries party to the Convention on Biological Diversity are enjoined to develop scientifically sound programs to conserve biodiversity, use the components sustainably, and share the benefits arising from such use fairly and equitably. This requirement is particularly important in the tropics, the region of the world where the highest levels of diversity are found, but also that where the flora and fauna are least well known and most under threat, where conservationists are relatively few and where the resources available for conservation activities are most limited. There is, therefore, an urgent need to clarify and enhance the methodologies that scientists use to classify, conserve, manage and use biodiversity. To be useful in developing countries, these methodologies must be appropriate, robust and cost effective.

Biodiversity is defined as the variability among all living organisms and the ecological complexes of which they are part. It includes not only diversity between species and ecosystems, but also genetic diversity within species. Crucial to the process of conservation and use of genetic diversity is the efficient management and appropriate analysis of the data it generates. These data are georeferenced, in that in the end they refer to samples from stands of wild or cultivated plants found growing at particular times *in specific places*.

Conservation data that are georeferenced can be linked to, and used in conjunction with, other georeferenced data (Table 1–1), from whatever source, e.g., on climate, soils, topography, human disturbance, and other aspects of the physical and biotic environment. This is important because the environment—through selection and isolation—plays a key role in structuring genetic diversity (Greene & Hart, 1999, see Chapter 2). The overall level, geographic distribution, and partitioning of variation into within- and among-populations components result from the processes of

Table 1–1. Use of georeferenced data on the target region and on the target taxon to carry out spatial analyses in support of different plant genetic resources conservation and use activities.

| Georeferenced data on target region | Georeferenced data on target taxon | Analyses | PGR activity |
|---|---|---|---|
| Thematic coverages (physical and biotic environment, socio-economic)<br>Interpolated climate surfaces | | Coverage overlay<br>Classification<br>Calculation of diversity statistics<br>Modeling<br>Search | Targeting of specific conditions or combinations of conditions<br>Identification of environmentally diverse areas (surrogate for genetic diversity)<br>Identification of areas threatened with genetic erosion<br>Production of aids for field exploration |
| | Passport data (incl. herbarium label data) | Overlay sites on coverages<br>Classification of conservation sites based on environmental data<br>Comparison of conservation sites with other sites | Checking passport data<br>Completing passport data<br>Modeling of potential distribution<br>Identification of ecotypes<br>Identification of underconserved areas (gap analysis)<br>Determination of seed storage and germination conditions<br>Design and zoning of in situ conservation areas<br>Designation of core collection<br>Determination of relative suitability of different regeneration and evaluation sites for different accessions and purposes |
| | Passport data<br>Characterization and evaluation | Calculation of diversity statistics<br>Classification of conservation sites based on characterization and evaluation data (and comparison with environmental classification)<br>Spatial autocorrelation (regional and population scales)<br>Differential systematics | Identification of high diversity areas<br>Identification of complementary conservation areas<br>Identification of ecotypes<br>Targeting of specific traits<br>Management and monitoring of in situ conservation areas<br>Designation of core collection |
| Remote sensing | | Environmental classification<br>Analysis of historical data<br>Seasonal time series analysis | Targeting specific habitats or crop-growing areas<br>Identification of areas threatened with genetic erosion<br>Timing of field exploration<br>Management and monitoring of in situ conservation areas |

mutation, selection (natural and artificial), gene flow and genetic drift, and thus depend on an interaction between various features of the biology of the plant, in particular its reproductive ecology, and its environment, including human activities.

In this chapter, we describe some methodologies that have been (or could be) applied to the analysis of georeferenced data by researchers working to enhance the efficiency and efficacy of the process of conservation and use of plant genetic resources. Many of the methodologies are based on the use of a Geographic Information System (GIS), and a brief introduction to GIS for the plant genetic resources worker is therefore also included. Rather than an exhaustive review of the subject, this is an attempt to structure discussion of the topic.

## A MODEL FOR CONSERVATION OF PLANT GENETIC RESOURCES

Maxted et al. (1997a) have developed a model for the process of conservation of plant genetic diversity that describes a series of linked, interdependent component activities, resulting in a set of products, including data, ultimately leading to the exploitation of genes, genotypes or populations within a gene pool for specific purposes (Fig. 1–1). The different stages of the process are briefly discussed in this section, focusing on the data that are needed for, and generated by, each task. In the following section we discuss how analysis of these and other georeferenced data can increase the efficiency of the different components of the process.

Selection of target taxon

Project commission

Ecogeographic survey/
Preliminary survey mission

Formulation of conservation objectives and strategy

Field exploration

Implementation of conservation strategy

↓                                                      ↓

*Ex situ* conservation methodologies          *In situ* conservation methodologies

↓                        ↓

Conservation products

Duplication and dissemination of conservation products

Characterization/evaluation of germplasm

Use of plant genetic resources

Use products

Fig. 1–1. Model of plant genetic resources conservation (Maxted et al., 1997a).

## Selection of Target Taxon

This will depend on such considerations as: current conservation status and threat of genetic erosion; taxonomic or genetic distinctiveness; biological, cultural and economic importance; cost, sustainability and probability of success of conservation; and the priorities of the agency concerned. Maxted et al. (1997c) discuss these in more detail.

## Ecogeographic Survey

This is a process of gathering, synthesizing and analyzing data on the ecology and geography of a region in which a target taxon occurs, as well as various aspects of its biology (taxonomy, ecology, distribution, phenology, reproductive biology, genetic diversity, seed storage behaviour, etc.) and ethnobotany (Maxted et al., 1995). At its core is the collation of the various types of sample-specific data associated with herbarium specimens (label data, observations of morphological traits) and germplasm accessions (passport, characterization and evaluation data). These are supplemented with more general information on target area and target taxon from various sources, and possibly with some preliminary field exploration. The ecogeographic survey brings together the basic information necessary to make sensible decisions about what, where and how to conserve genetic diversity within the target taxon.

## Conservation Strategy

The main product of the ecogeographic survey is a clear statement of the overall strategy to be followed to fulfill conservation objectives that are realistic and well defined. The strategy will ideally identify geographic areas and possibly specific stands requiring conservation interventions, and suggest what such interventions might be. Depending on species, region, resources available and other factors, this may involve the use of a mix of various kinds of both ex situ and in situ conservation methodologies. Each of these approaches has certain advantages and disadvantages, described further below (Bretting & Duvick, 1997; Maxted et al., 1997a).

## Field Exploration

The feasibility of the proposed conservation strategy will need to be verified. Where an ex situ approach is to be adopted, populations of the target taxon within the general target areas identified by the ecogeographic survey will have to be located, documented and sampled, and the germplasm brought back to the gene bank for conservation. If *in situ* conservation in a "genetic reserve" or on-farm is envisaged, populations of the target taxon will again need to be located and surveyed, and management and monitoring systems put in place. All this requires fieldwork. The location of the collecting site or *in situ* protected area is a fundamental component of the data collected during field exploration (passport data) because it helps document the environmental adaptation of the material and perhaps the level of isolation from other stands, and hence its potential usefulness and uniqueness. In the case of ex situ conservation, location of the collecting site may simply be recorded on field collecting forms as a locality name, but also as a more precise direction and distance from a landmark, as a grid reference on a specific map and/or as latitude and longitude. Usually, latitude and longitude are read off maps, but the increasing accessibility of Global Positioning System (GPS) receivers mean that more collectors are recording these data very accurately. In addition to its locality, collectors often record various ecologically significant features of the collecting site while in the field, including, for example, elevation, slope and aspect, vegetation type and perhaps soil type and pH. Data on some additional soil and climate descriptors can be added at a later stage, for example, after soil samples have been analyzed at the laboratory or the location of the collecting site has been compared with climate maps or data from the nearest meteorological stations obtained. However, it is quite rare at present for germplasm accessions to have such data.

## Implementation of Conservation Methodologies

Various options are available for both ex situ and in situ conservation (Bretting & Duvick, 1997; Maxted et al., 1997a). For example, for ex situ conservation, seeds and pollen may be maintained dry in cold stores, live plants in field gene banks or botanic gardens, and *in vitro* explants of various kinds in tissue culture or cryopreservation. The various complementary ex situ and in situ options available have different consequences for the amount and type of genetic diversity that can be maintained, the resources necessary for maintaining it and the ease with which it can be accessed and used. Also, different plants will be differentially amenable to the different options. For example, it is not now possible to conserve plants with recalcitrant seeds in ex situ seed gene banks long-term. Each conservation methodology can in turn be broken down into a series of component activities. For example, ex situ seed conservation includes the initial cleaning and preparation of samples for storage, possibly an initial seed multiplication, storage itself, viability monitoring, regeneration of accessions with low viability, etc. Again, each of these stages both generates and requires different kinds of data, which gene bank curators maintain in specialized documentation systems.

## Dissemination of Conservation Products

The products of conservation are germplasm, associated specimens such as herbarium and pest vouchers and microsymbiont samples, and—crucially—the data associated with all these, including sample-specific data and more general information such as collecting reports. These products will usually be stored in duplicate localities, and will generally be made available for exchange and use.

## Characterization/Evaluation

Description of germplasm is an important prerequisite for its use. Characterization refers to the process of making observations on traits that are mainly morphological, can be easily documented and are fairly consistently expressed in all environments. In contrast, evaluation is the measurement of characters that are of agronomic importance. These characters may be strongly influenced by the environment, so that replication over years and/or locations is needed for the data to be useful. Formal characterization and evaluation are often carried out by following standardized descriptor lists. These are published by the International Plant Genetic Resources Institute (IPGRI) for a wide range of gene pools (Perry & Bettencourt, 1995). Increasingly, DNA-based molecular techniques are being used to characterize germplasm, supplementing and complementing older morphological, physiological and biochemical approaches (Bretting & Widrlechner, 1995). Characterization and evaluation data on germplasm can be used to refine and extend the ecogeographic survey originally used to plan the collecting, as well as to structure the collection for easier use.

## Use of Germplasm

The use of conserved germplasm can take a variety of forms. It ranges from the breeder transferring useful alleles from different landraces into a new crop variety to the restoration of populations of a wild plant to an area where it has disappeared or of landraces to communities disrupted by civil strife or natural disaster. One of the advantages of in situ conservation is that it allows continued use of germplasm by local communities, although access to material by breeders and others may be more difficult. Use also includes more basic research, for example on seed storage behavior, taxonomic relationships or genetic diversity. Lack of adequate passport, characterization and evaluation information on large collections is often cited as a major constraint to their use (T. Hodgkin, 1997, personal communication).

## Documentation of Collections

We have seen that different kinds of data are generated at various stages during the process outlined in Fig. 1–1, and existing data used. Some of the basic categories into which the data generated by ex situ conservation have traditionally been grouped have already been mentioned, e.g., passport, characterization and evaluation. A full classification follows (based on the data categories used by SINGER, the Systemwide Information Network on Genetic Resources, see http://noc1.cgiar.org/grdata.htm).

### Identity and Origin of the Genetic Resources

**1. Passport (Collecting).** Data that are recorded at the site where the accession was collected (e.g., location of collecting site, latitude and longitude of collecting site, number of plants sampled, collecting source, etc.).

**2. Passport (Accession or Registration).** Data that describe the registration of accessions at the gene bank (e.g., accession number, acquisition date, type of germplasm).

## Characteristics of the Genetic Resources

**1. Characterization.** Data recorded during nonreplicated studies with the aim of initial description of traits that are not environmentally influenced.

**2. Evaluation.** Data recorded during replicated trials aimed at determining the true phenotype/genotype of material for traits that may be influenced by environmental conditions (e.g., protein content, yield, etc.). They may include characteristics of the site of characterization, or environmental data related to each trial.

## Distribution of the Genetic Resources

**1. Transfer.** Data that are recorded when material is distributed to collaborators (e.g., institute to which it was sent, accession number, date sent, etc.).

## Management of the Genetic Resources Collection

**1. Regeneration/Multiplication.** Data that are recorded when material is regenerated and multiplied (e.g., accession number, location of regeneration, type of pollination or isolation used, etc.).

**2. Inventory or Management.** Data that are recorded during the processes maintaining the genetic resources at a gene bank (e.g., location of sample in gene bank, seed viability, quantity of seed on hand, etc.).

The "collector's number" of a germplasm sample provides a link between the passport data, including the location of the collecting site, and the other data categories. Combined with the collector's name, this constitutes a unique identifier that is allocated to each germplasm sample in the field by the collector(s) and is never subsequently changed or dissociated from the sample. Thus, no matter what happens to a germplasm sample by way of transfer to another locality or sub-sampling for regeneration, duplication, use, etc., it should always be possible to trace its origin back to the specific locality where the stand from which it was collected is (or was) located. This is why the accession-specific data generated by the conservation and use process are described as "georeferenced".

Documentation systems for in situ conservation of plant genetic resources are not as well developed as for gene banks, but discussions of the topic can be found in Forde-Lloyd and Maxted (1997), Brockhaus and Oetmann (1996) and Bretting and Duvick (1997). However, because conservation in genetic reserves and on-farm implies an intensive and long-term involvement with a specific area of land, and with the people who live there, it will generate considerably more geo-referenced data than does ex situ conservation.

## USE OF GEOREFERENCED DATA IN PLANT GENETIC RESOURCES CONSERVATION AND USE

We have seen how the various components of the conservation and use process generate georeferenced data. They also make use of data of this kind generated by the process itself or available from outside sources. In this section, we discuss how these different kinds of georeferenced data can be analyzed to increase the efficiency of the following specific components of the conservation and use process: (i) ecogeographic surveying, (ii) field exploration, (iii) ex situ conservation—seed storage and regeneration, (iii) in situ conservation—design, management and monitoring of genetic reserves, (iv) evaluation, and (v) use of genetic resources.

However, we begin with a brief introduction to Geographic Information Systems. We will see that this technology provides an important tool for carrying out the kinds of analyses described below.

## Geographic Information Systems

A GIS is a database management system dedicated to the simultaneous handling of spatial data in graphics form and of related, logically attached, nonspatial data. For example, if the spatial data is the location of a collecting site, the associated attributes could be the names of the site, and other passport data about any germplasm samples that were collected there, as well as characterization, evaluation and management data. The main elements of a GIS are: (i) data input, verification and editing; (ii) data storage and database management; (iii) data manipulation and analysis; and (iv) data output (Guarino, 1995).

**Data input.** Spatial data can be entered into a GIS by digitizing, in which a map is mounted on a special electronic tablet and features are traced with a cursor or pointer, or by scanning, which involves generating a digital image of a map by moving an electronic sensor over its surface. Data from various kinds of remote sensing systems, from aerial photography to satellite imagery, also can be entered into a GIS. Attribute data are usually entered from a computer keyboard. Gene bank curators can enter passport, characterization and other data into the database of their documentation system and then import this into a GIS. Appropriate digital datasets for a given application may already exist. Some regional or global scale datasets are available from organizations such as FAO, UNEP/GRID, the International Soil Reference and Information Centre (ISRIC), the World Conservation Monitoring Centre (WCMC) and the international agricultural research institutes (IARC's) of the CGIAR. As an example, the spatial data holdings of the IARC's can be searched in the CGIAR Spatial Data Catalogue on the Web at www.grida.no/cgiar/htmls/mdindex.htm. One of the most commonly used medium-resolution base layer datasets is the *Digital Chart of the World* (DCW; ESRI, 1993), based on the 1:1 million Operational Navigational Charts and recently released on CD ROM.

**Data Storage.** There are two main types of GIS software, differing in how they store data (though "hybrid" systems are available). Vector-based systems store geographic data as points. Series of connected points make up lines, and lines enclosing an area make up polygons. In contrast, raster-based systems store data as grid cells, each representing a memory location in the computer. Lines are rows of grid cells and polygons are groups of adjacent grid cells. Vector systems require more computing power but less storage memory than do raster systems. They represent traditional map data better, because lines on the map remain lines, rather than becoming rows of grid cells. Vector systems will be best for some applications (archiving phenomenologically structured data such as topographic units or soil types, and for the highest quality output), raster systems for others (rapid overlay and combination of maps and spatial analysis) (Burrough, 1986).

**Data Manipulation.** Some standard GIS analytical capabilities include:

1. Geometric correction. The scale, projection, etc., of different maps may be changed to make them comparable.
2. Digital terrain model analysis. The elevation data (contours, spot elevations) on a topographical map may be used to produce maps of slope, aspect, intervisibility, shaded relief, etc.
3. Interpolation. Point data may be used to create isopleth (equal-value contour) maps.
4. Overlay analysis. Different maps of the same area may be combined to produce a new map, e.g., maps of slope, soil, wind speed and vegetation cover may be overlaid to synthesize a map of soil erosion risk.
5. Proximity analysis. Buffers or windows may be generated around features.
6. Computation of statistics. Means, counts, lengths, areas etc. may be calculated for different features.
7. Location. Entities having defined sets of attributes may be located.

The spatial processing system and database management system of a GIS thus allow one to bring together diverse datasets, make them compatible among themselves, and analyze and combine them in different ways.

**Data Output.** The results of these kinds of analyses must be displayed for them to be of use, and the ability to produce high-quality hard copies of the results of analyses is an important feature of GIS software and hardware. The software usually allows such manipulations as selecting particular areas or layers of a map for output, scale change, color change, etc.

Most of the analytical and display capabilities of a GIS can find application in the field of plant genetic resources conservation. However, digital terrain models are of particular importance. These have been used to increase the efficiency of interpolation between meteorological stations and thus produce accurate, high-resolution, continent-scale surfaces for various temperature and precipitation variables (e.g., Hutchinson et al., 1996). Such interpolated climate surfaces are potentially of great use in plant genetic resources conservation and use work and will be alluded to on a number of occasions in the following sections, which discuss the analysis of georeferenced data in the context of various specific components of the conservation process.

## Ecogeographic Survey

The essential first step in the development of a comprehensive strategy for the conservation and use of plant genetic resources is some form of ecogeographic survey. It is only on the basis of an understanding of the taxonomy, genetic diversity, geographic distribution, ecological adaptation and ethnobotany of a plant group—and of the geography, ecology, climate and the human setting of potential target regions—that sensible conservation decisions can be made. Key issues that this kind of analysis may help elucidate include: when, where and how to collect germplasm; and where genetic reserves might best be placed and how they need to be monitored and managed.

The process of carrying out an ecogeographic survey is described in detail by Maxted et al. (1995), who break it down into a number of distinct stages. The key activities are collating sample-specific data from herbarium specimens and from germplasm accessions and more general information from experts, the published literature, maps, databases, etc. A main product of the ecogeographic survey will be maps showing areas that: (i) should contain specific target traits, taxon or habitats; (ii) are highly diverse (in terms of the environment, taxonomically or genetically); (iii) are different from each other environmentally, genetically or taxonomically; (iv) are missing or under-represented in conservation efforts; and (v) may be threatened with genetic erosion. Some approaches to identifying such areas are discussed in turn below.

## Areas Likely to Contain Target Germplasm

Distribution data are often scanty and patchy. It would therefore be useful to be able to identify areas where a species has perhaps not actually been recorded (because the area has not been visited by botanists or germplasm collectors) but where it might still be expected to be found on the basis of what little is known of its distribution. Software such as BIOCLIM (Busby, 1991), DOMAIN (Carpenter et al., 1993) and that used by Jones et al. (1997) use climate interpolation surfaces (and potentially other environmental data layers, such as soil maps) to estimate conditions at each of a set of sites where a species has been recorded (e.g., collecting sites of germplasm or herbarium specimens) given their latitude, longitude and altitude. They derive an "envelope" for the set of localities and then display all other localities that show given levels of similarity to the collecting sites. Afonin and Greene (1999, see Chapter 6) describe a similar methodology.

The difference between these different approaches is that the BIOCLIM and DOMAIN software express the habitat envelops in terms of raw variables such as annual mean temperature and annual precipitation. In contrast, the technique of Jones et al. (1997) uses multivariate statistics first, in order to determine what linear combination of variables best summarizes the variation in the data, and then uses the transformed variables to characterize the habitat envelope (P. Jones, 1997, personal communication). Model-based approaches (e.g., Walker, 1990; Stockwell & Noble, 1992) also have been used to investigate the so-called "potential" distribution of a species. These techniques seem thus far only to have been applied at the species level. However, there is no reason why they could not be tried on intraspecific entities such as botanical varieties, groups of similar accessions (e.g., based on multivariate analysis of morphological or molecular characterization), or even individual landraces (i.e., traditionally recognized crop entities). Incidentally, Jones et al. (1997) make the point that their technique also can be used to investigate whether a set of accessions of a given species actually consists of different ecotypes or gene pools with distinct climatic envelopes. This is useful information to the conservationist because if such ecotypes exist, they would best be targeted separately for conservation to ensure that as much as possible of the range of genetic diversity within the species is covered.

Users often require that material with specific adaptations be targeted for conservation. A GIS can be used to manipulate and superimpose appropriate thematic environmental coverages from disparate sources to identify areas where material with the required adaptation might be expected to occur. Afonin and Greene (1999, see Chapter 6) illustrate an example of collecting forage germplasm adapted to arid conditions in the Caucasus. Pollak and Corbett (1993) used monthly climate averages to define different zones of maize (*Zea mays* L.) adaptation in northern Latin America. By comparing the results of such an analysis with their target environments, breeders would be able to say which areas they consider high priorities for germplasm collecting. The Spatial Characterization Tool (SCT) is one example of a GIS application that can carry out this kind of search (Corbett & O'Brien, 1997). The SCT will eventually incorporate data for Africa, Latin America and parts of Southeast Asia at resolutions from 30 arc-seconds to 3 arc-minutes (5.4 km at the equator), including climate, topography, land cover, demographic and soils information, as well as ancillary information on settlements, political units, bodies of water and major watersheds. The climatic data include long-term monthly normals for precipitation, evapotranspiration, the ratio of precipitation to potential evapotranspiration (P/PE) and maximum, minimum and mean temperature, as well as various models of the growing and dry seasons.

If collection and characterization/evaluation of germplasm have already been carried out, geostatistical tools such as variograms (autocorrelation or spatial covariance function) and surface interpolation can be used to describe the spatial pattern of variation in genetic, morphological and agronomic traits (or combinations of traits) among populations on a regional scale. The effects of selection on a broad scale can then be explored by seeking correlations with environmental and other factors, and disentangled from the effects of isolation by distance and microenvironmental variation (e.g., Monestiez et al., 1994). Another approach is to use differential systematics (Kirkpatrick, 1974), which involves the combination of different character contour maps into a single map of a "systematic function," the ridges of which reveal areas where maximum change over distance is occurring, which have been called genetic boundaries (see also Monmonier, 1973; Pigliucci & Barbujani, 1991). Such analyses can guide the user to areas where target traits are prevalent, but also help the conservationist identify areas that are relatively homogeneous but different from each other for the characters being studied. This is discussed further in the section on complementarity.

Finally, remote sensing data can be useful in locating areas of potential interest, especially if they are remote and fragmented. Different target vegetation and land use types can often be recognized in aerial photographs and satellite imagery. For example, Landsat Thematic Mapper (TM) imagery was used by Veitch et al. (1995) to map heathland fragments in southern England, and by Hart et al. (1996) to locate meadows in the western Caucasus. Isolated areas of cultivation also can be identified, and in some cases the types of crops grown recognized.

## Highly Diverse Areas

Some geographic areas show greater taxonomic and/or genetic diversity than others. Diversity studies usually begin by dividing the target area (or strata within the target area, e.g., climate zones) into areas of equal size (to reduce the species-area effects on diversity and rarity measures). These are often squares in a grid pattern, but geometric, political or economic spatial units have been used (e.g., see references in Csuiti et al., 1997), as well as point-centered approaches (M. Grum, 1997, personal communication). Some measure of diversity is calculated for each spatial unit, which can then be compared with each other. For example, Nabhan (1991) used presence/absence of species to investigate patterns in the taxonomic diversity of wild *Phaseolus* in different grid squares in the Sierra Madre, Mexico.

Such analyses of taxonomic diversity can be refined further. Measures of diversity can be corrected for differences in sampling intensity among the different grid squares by various parametric and nonparametric methods (Gaston, 1996). Such a correction has been attempted for data on wild potato (*Solanum tuberosum* L.) distribution in Bolivia (R. Hijmans, 1997, personal communication). An approach to correction is yet to be explored. This involves defining the potential distribution of species (or other units) using procedures such as that of Jones et al. (1997), and then calculating the diversity occurring in different areas using this rather than observed distributions.

Another possible refinement may be necessitated by the fact that two grid squares may have equal numbers of taxa (i.e., equal diversity according to one measure), but the ones in one square may be very similar to each other (i.e., closely related) while those in the other may be very different. Other things being equal, the second grid square would be the higher conservation priority (Humphries et al., 1995). The procedure described by Vane-Wright et al. (1991) and available in their WORLDMAP software allows a taxonomic diversity measure to be weighted for the distinctness of taxa, calculated from a phylogeny based on the presence/absence of characters. Measures of diversity based on morphological characters can be used at the infraspecific level. Thus, Pickersgill (1984) used characterization data to calculate the morphological diversity (using Shannon-Weaver diversity index) shown by accessions of cultivated *Capsicum* spp. in different grid squares within Central and South America.

Numerous studies have attempted to use diversity in different environmental parameters as a surrogate for taxonomic diversity (Gaston, 1996). One example among many is the work of Miller (1986), who showed that variation in elevation, calculated by GIS for each of a number of polygons in the southern Appalachian region, is a useful predictor of the richness of rare species in those polygons. These kinds of studies have had varying levels of success, and do not seem to have been applied at the intraspecific level as yet. However, a study is investigating the relationship between environmental and human diversity and genetic diversity in cultivated peanut (*Arachis hypogaea* L.) in Ecuador and Guatemala (D.E. Williams & K.A. Williams, 1997, personal communication). Afonin and Greene (1999, see Chapter 6) also used environmental diversity to guide their forage collecting.

## Dissimilar Areas

It is not enough to simply target areas that are highly diverse to maximize the amount of diversity protected for a given amount of effort, because all the areas might actually contain the *same* diversity (as well as the same amount of diversity). One approach to the optimal targeting of conservation effort is to use multivariate statistics to classify or ordinate spatial units according to the species found there, or on the basis of characterization/evaluation data when dealing with a single crop or species. Indeed, this can be done in terms of environmental conditions too, and environmental classification is a common GIS application (Pollak & Corbett, 1993; Booth et al., 1989). Spatial units can then be sampled separately in a stratified manner from each distinct cluster of similar units.

More sophisticated methods also are available. Iterative procedures such as that described by Rebelo and Sigfried (1992), can be used to choose the smallest number of spatial units such that each species, morphotype, etc., will be present in at least one unit in the set (or two, three, etc.). A recent study of 19 different techniques (Csuti et al., 1997) found that this and various other heuristic techniques can be very efficient at solving biodiversity conservation site selection problems, but (except when dealing with large, complicated data sets), recommends the use of an approach to linear programming called a "branch-and-bound" algorithm.

## Under-Conserved Areas

The WORLDMAP software mentioned earlier has a facility that allows the user to select grid squares so that a subsequent run identifies those grid squares that are complementary to the selected ones. If the selected areas are existing protected areas, this becomes the kind of study often referred to as "gap analysis". The concept and some applications are discussed in detail by Scott et al. (1993). The process involves the use of "digital map overlays in a GIS to identify individual species, species-rich areas, and vegetation types that are not represented or under-represented in existing biodiversity management areas". Although this has not been considered in the published work on gap analysis, the "existing biodiversity management areas" could just as well be areas where germplasm collecting has already been adequately carried out as protected areas of different kinds. Identifying ecogeographical gaps in existing ex situ germplasm collections in this way is increasingly important

## Threatened Areas

Satellite imagery and other remote sensing systems can provide information on long-term trends in vegetation and land use change in an area. Deforestation and desertification in particular have been documented at a variety of scales using remote sensing data stretching back over many years (e.g., Skole & Tucker, 1993; Gastellu-Etchegorry et al., 1993). Working in Côte d'Ivoire, Chatelain et al. (1996) have documented a "deforestation front" of forest fragments that in some cases is approaching or actually encroaching on protected areas. This kind of analysis can identify areas that have experienced changes, but also assist in predicting what areas are most at risk.

Mathematical modeling also can do this. For example, data on topography, soil, land use and precipitation have been combined in mathematical models to predict the risk of soil erosion on scales ranging from local to global. Work by Fujisaka et al. (1996) at two sites in the Brazilian Amazon combining GIS with farmer interviews has investigated the influence on deforestation of distance to roads, wet-season access, land tenure and parcel size. These findings could be incorporated into a predictive model of deforestation on a wider scale. In a similar way, data on demography, economic development, accessibility, potential for irrigation and agroclimatic suitability for cash crops or improved varieties are used to develop a model for the risk of genetic erosion in peanut in Ecuador and Guatemala (D.E. Williams & K.A. Williams, 1997, personal communication).

## Field Exploration

Fieldwork is needed to refine the desk-based ecogeographic survey and implement the conservation strategy developed on the basis of its results, whether ex situ or in situ. In this section, we discuss how GIS can assist in the development of field aids, in timing field visits and in documenting any germplasm collected.

### Field Aids

Various kinds of documentation are required in the field, but maps are especially important. Greene et al. (1999a) describe the development of GIS mapping products for germplasm collecting to support a joint U.S./Russian forage collecting program in the Caucasus. This included a number of moisture maps and temperature maps synthesizing some 60 monthly climate variables at 500-m Universal Transverse Mercator (UTM) grid cell resolution. A soil map also was synthesized from four different map series. These maps allowed the identification of ecogeographic gradients in the field, the slope of which influenced sampling frequency in this case, but which also could have been used to delimit potential in situ conservation areas. The maps also were used to monitor which combinations of conditions were being adequately sampled and which were not.

### Timing

Satellite imagery can provide data on vegetation development with very short lag-times. Use of such data might allow collectors to be very much more precise in timing their visit. This is particularly important in the arid and semiarid tropics, where rainfall, and therefore vegetation development, is unpredictable in both space and time. For example, Meteosat and NOAA/AVHRR data on rainfall and the state of vegetation (as measured by the Normalized Difference Vegetation Index, or NDVI), though fairly low in resolution, can be analyzed to allow surveillance of the state of crops and vegetation. Justice et al. (1987) describe the annual course of NDVI in a variety of East African vegetation types, and how this measure relates to the phenology of rainfall and plant growth. One could conceivably read off the latitudes and longitudes of a set of potential target collecting areas from such satellite imagery and use a GPS receiver to locate them in the field a matter of days later. The U.S./Russian forage collecting program used Landsat Multi-Spectral Scanner (MSS) and TM imagery both to locate the primary target meadow patches and to evaluate the stability of their phenology across years (Hart et al., 1996).

## Completing Passport Data

Collectors have long appreciated the importance of providing herbaria and genebanks with material that is as fully documented as possible, but the fact remains that, according to the SINGER database, germplasm accessions with latitude and longitude data account for only about 20% of CGIAR collections (elevation data is somewhat more common). Accessions with associated climatic data are even fewer. The GIS technology can help the collector and gene bank curator in obtaining fuller and more accurate passport data. The first step is to obtain latitude and longitude data for accessions where only a locality name is available, by referring to the collector's notebooks and maps, gazetteers, etc., as necessary. Such data as altitude, major soil type, land use and vegetation, if missing, can then be estimated by overlaying the locations of collecting sites on different digitized base maps of the appropriate scale. Interpolated climate surfaces can then be used to characterize conditions at each collecting site given its latitude, longitude and altitude. Steiner and Greene (1996) describe this process as "retro-classification" or "retro-characterization" of accessions and give an example using the U.S. National Plant Germplasm System (NPGS) *Lotus* collection.

## Ex Situ Conservation

Once germplasm has been obtained as a result of field work, the material must be maintained in ex situ gene banks. This process consists of a number of activities. Two of these, seed storage and regeneration, are particularly reliant on geo-referenced data on the site of origin of the material.

### Seed Storage

Data on environmental conditions at the collecting site may be important at various stages of the process of seed storage. In rice (*Orya sativa* L.), seed production environment has been shown to have an effect on the development of desiccation tolerance to low moisture contents, and therefore on longevity in gene banks (Ellis et al., 1993). Monitoring the viability of conserved seed depends on regular germination testing of subsamples of each accession. A compendium of germination protocols for a wide range of species has been published (Ellis et al., 1985), but there is evidence of intraspecific variation in germination requirements linked to environmental factors (e.g., Gutterman & Nevo, 1994). Climatic and other data on collecting sites, obtained retroactively if necessary, may therefore be useful in fine-tuning germination protocols, in particular for dormancy-breaking. The suggestion also has been made that seed storage behavior is associated with ecology at the interspecific level, with recalcitrant species tending to originate from moist ecosystems. For recalcitrant and intermediate species, there may be an association between optimum seed storage temperature and the minimum air temperature at which the plants can survive without chilling injury (Hong & Ellis, 1996). Climate surfaces might be used to determine minimum air temperatures within the distributions of different recalcitrant and intermediate species with a view to determining optimum seed storage conditions.

## Regeneration

Regeneration is an important component of ex situ conservation. It is the process of identifying those accessions in a collection with inadequate quality or quantity of seed and producing from them a new seed sample of adequate size, maximum seed quality and as nearly as possible the same genetic composition of the original (Sackville Hamilton & Chorlton, 1997). One of the key decisions that has to be made by gene bank managers in planning their regeneration program is where to carry out this work. The environment of the regeneration site must not only provide suitable conditions for reliable flowering and seed production but also any necessary triggers for the different stages of plant development. Any necessary pollinators also must be present (except in case of controlled pollination in cages, etc.). If possible, pests and diseases must not be prevalent and populations of species related to the one(s) being regenerated should not be found nearby.

The ideal environmental conditions in which to regenerate an accession will not necessarily be exactly the same as those where it was collected, because plants are often collected from suboptimal seed production environments. Nevertheless, data on collecting sites derived from climate surfaces would still be useful in a regeneration program. In conjunction with climate surfaces, climatic characterization of each accession also might be used to determine the locations of the minimum number of sites in a country or region where conditions would be suitable for regenerating the collection as a whole. If the choice of locations for regeneration is restricted, as it will be in most cases, such data can be used to estimate the relative suitability of different available sites for the regeneration of different accessions. They also can help to determine the unique needs of individual accessions so that growing procedures can be customized in suboptimal sites.

## Evaluation

A similar problem of matching accessions to sites exists when deciding whether (and which) existing sites are suitable for evaluating which accessions for various different traits. This tends to be the domain of breeders and other users rather than plant genetic resources conservationists, but some brief observations are in order. Breeders usually use both representative and stress locations for their germplasm evaluation. Classifications of testing locations, such as those of Pollak and Pham (1989), can be compared with crop climate classifications to determine the representativeness of different sites and their suitability for testing for different traits. Time-series climatic data can be used to determine the frequency with which a given location produces a particular stress (Chapman & Barrato, 1996). Once a site has been chosen, GIS can be used to determine the extent of the region where similar conditions apply, and hence the relevance of the evaluation work (Chapman & Barrato, 1996).

## In Situ Conservation

As has already been pointed out, in situ conservation will probably both generate and require georeferenced data to a much greater extent than ex situ conservation. In this section, we briefly discuss some examples of how such data can be used in the design, management and monitoring of areas where plant genetic resources are conserved in situ, either in genetic reserves or on-farm.

## Design

There is an extensive literature on the design of protected areas, dealing with such questions as optimal size and shape, zonation, networking etc. (e.g., Given, 1994), though, as Hawkes et al. (1997) point out, discussion has largely centered on the requirements of habitat conservation, especially for wild animal species such as large mammals and birds. These are spatial problems, and spatial analysis and GIS have been applied to them. For example, Howard (1996) discusses how spatial information on species richness, distribution and abundance of an endangered species, disturbance, and distribution of timber resources within a forest can be used to develop a zoning plan, including different use areas, buffer zones and a core. This type of application presents the challenge of integrating indigenous knowledge and demographic, socioeconomic and other data on the human population with data on the physical environment and on the target taxon. Fox et al. (1996), for example, used GIS to map areas where the objectives of protecting red pandas and those of meeting the claims and grazing needs of communities living in the area either coincide or come into conflict. The need for such integration will be especially great in the case of on-farm conservation of crop genetic resources. A preliminary attempt at this is described by Salick et al. (1997) for cassava (*Manihot esculenta* Crantz) among the Amuesha people in the Peruvian Amazon.

## Management

There are numerous examples of the maintenance, manipulation and analysis of georeferenced data in a GIS for better reserve planning and management. At the habitat level, two relevant examples are Kessell's (1990) GIS-based decision support system for land management in fire-prone rural land in Australia and Legg's (1995) database on the tropical forests of the Knuckles Range, Sri Lanka. At the level of the population, basic demographic and genetic structure data on a plant within a prospective genetic reserve can be used to carry out procedures such as transition matrix sensitivity analysis (Silvertown & Lovett Doust, 1993) or population viability analysis (Menges, 1991). These can in turn be used to model and predict the effect of different management interventions on numbers and genetic diversity. For example, Liu et al. (1995) used a spatial population model in a GIS to simulate the potential effects of different forest management plans on a bird species.

## Monitoring

Just as ex situ conservation requires tracking seed quantity and periodic measurement of seed viability, in situ conservation also requires regular monitoring (Maxted et al., 1997b). Monitoring requires decisions about the variables to document (abundance, demographic parameters, genetic diversity, etc.), and where, how, how much, and how often to sample. In any case, in contrast to ex situ conservation, it will often be individuals (or groups of individuals in permanent quadrats), rather than populations, that will need to be documented, adding considerably to the volume of georeferenced data generated.

Regarding where to sample, georeferenced data on the physical and biotic environment of the site can be used to produce classifications to serve as the basis for stratified sampling. Documenting the location of individuals will be necessary for studies of population dynamics (Harper, 1977), including the estimation of such important parameters for conservation purposes as transition probabilities between different life history stages, which are necessary for sensitivity analysis and population viability analysis. Geostatistical tools such as spatial autocorrelation analysis can be used on characterization (including biochemical and molecular) data from individual plants for studies of the spatial genetic structure of populations, selection, gene flow, hybridization, etc., also very relevant to the conservation of genetic diversity, particularly in situ (e.g., Waser, 1987; Epperson & Allard, 1989; Epperson, 1993).

## Use of Genetic Resources

Conservation will be afforded inadequate resources unless the benefits from investment in germplasm maintenance, which flow from thorough and effective use, can be demonstrated. However, there are various bottlenecks to increased use. Apart from institutional issues such as restricted access policy, inadequate linkages between gene banks and users and the limited capacity of breeding programs to absorb new material, the most often mentioned problems are lack of data on accessions, the presence of genotype × environment interactions (making evaluation outside the environment of use irrelevant) and the large size of collections (T. Hodgkin, personal communication).

### Inadequate Data

We have already mentioned in the section on field exploration how GIS can assist in solving the problem of incomplete or inadequate passport data by "retro-characterization" of collecting sites. In addition to being incomplete, passport data also are sometimes inaccurate. A place name could refer to a number of different localities, or there could have been confusion over different coordinate systems or errors in entering data (Chapman & Busby, 1994). Data exploration using a GIS can quickly reveal obvious errors in latitude/longitude data resulting in accessions falling into the sea, a lake or the wrong country or administrative region (R. Hijmans, 1998, personal communication; Maxted et al., 1995). Obvious geographic outliers also may be suspected. Chapman and Busby (1994) describe an innovative method of spotting outliers that involves plotting the climate profile of each specimen and looking for entries that are out of step with the rest. These could represent distinct taxa or ecotypes, as suggested by Jones et al. (1997) and alluded to in the section on ecogeographic surveys, or simply errors in the data.

### Targeting Material

The kind of detailed retro-characterization of the collecting site described by Steiner and Greene (1996) may assist the gene bank curator in identifying suitable sites for regenerating and evaluating different accessions, as already discussed. In addition, it can guide the use of the material, by focusing the attention of users on the most promising material for their different specific purposes. Crop-environment classifications have been mentioned in the context of ecogeographic surveys because they can be used to identify geographic areas where target material may be located. Of course, if a collection has already been made, such classifications, applied retroactively to accessions, can likewise be used to identify material most likely to be of use for specific purposes. For example, although it has been developed in the context of introduction of tree species, the climate matching programs demonstrated by Booth (1990) for various countries may be valuable tools for targeting the introduction of germplasm to specific areas for evaluation and use based on conditions where it was collected. Chapman and Taba (quoted by Chapman & Barreto, 1996) overlayed germplasm locations on soil maps to identify candidates from an extensive maize collection that might be adapted to alkaline soils. However, Beebe et al. (1997) made the point that for landraces, which farmers have been moving around for centuries, the environment of collection may not correspond to the environment where the material evolved. A GIS can be used to calculate, for a region of given size surrounding each collecting site, measures of the variability or prevalence of target environments, which might be a better pointer to landrace adaptation than conditions at the collecting site itself.

## Core collections

Another way the problem of using large collections has been approached, in addition to better targeting of material, is by using a specially selected subset, called a "core collection", as a gateway into the collection as a whole. According to Brown (1993), "A core collection is a selected and limited set of accessions derived from an existing germplasm collection, chosen to represent the genetic spectrum in the whole collection and including as much as possible of its genetic diversity." Establishing a core collection essentially involves classifying accessions into groups that are relatively genetically homogeneous internally but different from each other. A certain number or proportion of the accessions are then selected from each group. Both classification and selection can be carried out in a variety of ways. The case of the *Phaseolus vulgaris* core collection described by Tohme et al. (1995) illustrates one use of georeferenced data. First, regions were prioritized on the basis of the history of the crop. Then, interpolated surfaces for four parameters (length of growing season, photoperiod, soil type and moisture regime) were used to define 54 distinct environments, and each 10-min grid cell was assigned to one of these classes. Passport data were then used to match each landrace accession to an environmental class. Finally, accessions in each environmental class were stratified according to characterization data (growth habit and grain color and size) and selections made at random from within each stratum within each environmental class.

Brown (1993) points out that it is in the area of evaluation and use that the core collection may have most to offer. It provides a limited set of accessions (perhaps 10% of the total) that can be used to evaluate traits that are expensive or time consuming to measure, for example complex yield and quality traits and general combining ability with local germplasm. These studies can identify material in the rest of the collection that might repay investigation. Different workers have used passport, characterization (morphological, molecular, etc.) or evaluation data, often in different, usually hierarchical, combinations, to classify accessions. For example, compare Tohme et al. (1995) and Hamon et al. (1995). However, the importance of location data is widely acknowledged and agroclimatic conditions at the collecting site, as in the *Phaseolus* case described above, are often one of the more important criteria for classification.

## Prospects

Table 1–1 summarizes how georeferenced data on the target region and on the target taxon from different sources may be combined to assist various activities in the conservation and use of plant genetic resources. What lies ahead in this field? First, the data will surely get better. It will hopefully be increasingly rare that herbarium and germplasm collectors do not record latitude, longitude and elevation in the field. The wider availability of GPS receivers will help here, but also collectors are better trained: the importance of high-quality, standardized passport data must continue to be stressed when teaching plant genetic resources conservation. At the same time, curators of collections with missing or inconsistent data are trying to rectify the situation. This is certainly true of the CGIAR centers and of the U.S. National Plant Germplasm System, probably the two largest holders of genetic resources collections (e.g., Greene & Hart, 1996). Similarly, climate interpolation surfaces and other environmental digital data sets, as well as socioeconomic data sets, will improve in geographical coverage, resolution and accuracy. They also will become more easily available.

However, if the use of GIS is to spread significantly beyond the IARCs and the national plant genetic resources conservation and use programs of a few countries, it is not sufficient that there be more, better data available. Training of plant genetic resources workers in the analysis of their georeferenced data by GIS will be necessary. Linked to this is the need for cheap, easy-to-use, specialized software tools. This will probably require national plant genetic resources programs to enter into partnerships with universities and public sector companies, either within the country or abroad, for the provision of hardware, software and training. The IARCs, and national programs with GIS experience, can play an important role in promoting such international, multidisciplinary linkages to develop appropriate technology packages (see Afonin & Greene, 1999, see Chapter 6). A glimpse into the future is perhaps afforded by the International Crop Information System being developed by CIMMYT and IRRI (Fox & Skovmand, 1996). This will include linked modules for the management of gene bank, pedigree, characterization/evaluation, and environmental data. The integration of a GIS into the system will allow some of the kinds of manipulation of georeferenced data that we have been discussing. Chapman and Barreto (1996) have suggested that as the linkage between genetic resources and environmental data is explored, "geographic trait loci" may be identified, where particular parts of the genome are linked to adaptation to different conditions. The Spatial Characterization Tool is a step in the right direction.

As GIS is used more and more to analyze georeferenced conservation data, some of the technical problems discussed by a number of the contributors to this volume, in particular by Hart (1999, see Chapter 4), will begin to be resolved, and others will arise. The important thing is that the technology should be adequately tested. Some impetus for this may well come from the increasing importance accorded by plant genetic resources conservation programs in many countries to in situ and on-farm conservation. In this type of conservation, it is important to collect and analyze data not just on populations, the usual case in ex situ conservation, but sometimes also on individual plants, allowing studies of demography and population dynamics, of spatial genetic structure within populations, of mating systems, of gene flow, etc. Data collection will need to be repeated regularly for monitoring purposes. Different areas within the prospective genetic reserve may be used and managed in different ways and be subject to different pressures. All this will produce more georeferenced data than has been the case in ex situ conservation, and plant genetic resources conservationists will need to borrow extensively from among the spatial analytical methods developed by land managers, protected-area conservationists, ecologists, population geneticists, etc.

The idea of adapting techniques from other disciplines is one that has been alluded to a number of times in this chapter. This is perhaps a useful point to reiterate in conclusion. The data and methods will be increasingly available. The challenge for plant genetic resources workers is to put them to work to solve their particular problems. This will require institutional linkages at both national and international level. It also will require better awareness at both the technical and the decision-making level within national and regional programs of the enormous possibilities that GIS technology offers for the conservation and sustainable use of plant genetic resources.

## REFERENCES

Afonin, A., and S.L. Greene. 1999. Germplasm collecting using modern geographic information technologies: Directions being explored by the N.I. Vavilov Institute of Plant Industry. p. 75–85. In S.L. Greene and T.C. Hart (ed.) Linking genetic resources and geography: Emerging strategies for conserving and using crop biodiversity. CSSA Spec. Publ. 27. ASA and CSSA, Madison, WI.

Beebe, S., J. Lynch, N. Galwey, J. Tohme, and I. Ochoa. 1997. A geographical approach to identify phosphorus-efficient genotypes among landraces of common bean. Euphytica 95:325–336.

Bretting, P.K., and M.P. Widrlechner. 1995. Genetic markers and plant genetic resource management. Plant Breed. Rev. 13:11–86.

Bretting, P.K., and D.N. Duvick. 1997. Dynamic conservation of plant genetic resources. Adv. Agron. 61:1–51.

Booth, T. 1990. Mapping regions climatically suitable for particular tree species at the global scale. For. Ecol. Manage. 36:47–60.

Booth, T.H., S.D. Searle, and D.J. Boland. 1989. Bioclimatic analysis to assist provenance selection for trials. New Forest 3:225–234.

Brockhaus, R., and A. Oetmann. 1996. Aspects of the documentation of in situ conservation measures of genetic resources. Plant Gen. Resour. Newsl. 108:1–16.

Brown, A.H.D. 1993. The core collection at the crossroads. p. 3–19. In T. Hodgkin et al. (ed.) Core collections of plant genetic resources. John Wiley & Sons, Chichester, UK.

Burrough, P.A. 1986. Principles of Geographic Information Systems for land resources assessment. Clarendon Press, Oxford, UK.

Busby, J.R. 1991. BIOCLIM–a bioclimate prediction system. p. 64–68. In C.R. Margules, and M.P. Austin (ed.) Nature conservation: Cost effective biological surveys and data analysis. CSIRO, Melbourne, Australia.

Carpenter, G., A.N. Gillison, and J. Winter. 1993. DOMAIN: A flexible modelling procedure for mapping potential distributions of plants and animals. Biodivers. Conserv. 2:667–680.

Chapman, A.D., and J.R. Busby. 1994. Linking plant species information to continental biodiversity inventory, climate modelling and environmental monitoring. p. 179–195. *In* R.I. Miller (ed.) Mapping the diversity of nature. Chapman & Hall, London, UK.

Chapman, S.C., and H.J. Barreto 1996. Using simulation models and spatial databases to improve the efficiency of plant breeding programs. p. 563–587. *In* M. Cooper, and G.L. Hammer (ed.) Plant adaptation and crop improvement. CAB Int., Wallingford, UK.

Chatelain, C., L. Gautier, and R. Spichiger. 1996. A recent history of forest fragmentation in southwestern Ivory Coast. Biodivers. Conserv. 5:37–53.

Corbett, J.D., and R.F. O'Brien. 1997. The spatial characterization tool–Africa. Vers. 1.0. Texas Agric. Exp. Stn. Blackland Res. Center Rep. 97– 03.

Csuti, B., S. Polasky, P.H. Williams, R.L. Pressey, J.D. Camm, M. Kershaw, A.R. Kiester, B. Downs, R. Hamilton, M. Huso, and K. Sahr. 1997. A comparison of reserve selection algorithms using data on terrestrial vertebrates in Oregon. Biol. Conserv. 80:83–97.

Ellis, R.H., T.D. Hong, and E.H. Roberts. 1985. Handbook of seed technology for genebanks. Vol. 2. Handbooks for Genebanks No. 3. Int. Board Plant Genet. Resour., Rome, Italy.

Ellis, R.H., T.D. Hong, and M.T. Jackson. 1993. Seed production environment, time of harvest, and the potential longevity of three cultivars of rice (*Oryza sativa* L.). Ann. Bot. 72:583–590.

Environmental Systems Research Institute, Incorporated. 1993. Digital map of the world. ESRI, Redlands, CA.

Epperson, B.K. 1993. Recent advances in correlation studies of spatial patterns of genetic variation. Evol. Biol. 27:95–155.

Epperson, B.K., and R.W. Allard. 1989. Spatial autocorrelation analysis of the distribution of genotypes within populations of lodgepole pine. Genetics 121:369–377.

Forde-Lloyd, B.V., and N. Maxted. 1997. Genetic conservation information management. p.176–191. *In* N. Maxted et al. (ed.) Plant genetic conservation: The *in situ* approach. Chapman & Hall, London, UK.

Fox, J., P. Yonzon, and N. Podger. 1996. Mapping conflicts between biodiversity and human needs in Langtang National Park, Nepal. Conserv. Biol. 10:562–569.

Fox, P.N., and B. Skovmand. 1996. The International Crop Information System (ICIS)—connects genebank to breeder to farmers's field. p. 317–326. *In* M. Cooper and G.L. Hammer (ed.) Plant adaptation and crop improvement. CAB Int., Wallingford, UK.

Fujisaka, S., N. Thomas, and E. Crawford. 1996. Deforestation in two Brazilian Amazon colonies: Analysis combining farmer interviews and GIS. *In* Proc. 3rd Int. Conf./Workshop on Integrating GIS and Environ. Model., Santa Fe, NM. 21–26 January. Natl. Center for Geograph. Inform. Anal., Santa Barbara, CA (http://www.ncgia.ucsb.edu/conf/SANTA_FE_CD-ROM/main.html).

Gastellu-Etchegorry, J.P., C. Estreguil, E. Mougin, and Y. Laumonier. 1993. A GIS based methodology for small scale monitoring of tropical forests—a case study in Sumatra. Int. J. Remote Sens. 14:2349–2368.

Gaston, K.J. 1996. Species richness: Measure and measurement. p. 77– 113. *In* K.J. Gaston (ed.) Biodiversity: A biology of numbers and difference. Blackwell Sci., Oxford, UK.

Given, D.R. 1994. Principles and practice of plant conservation. Chapman & Hall, London, UK.

Greene, S.L., and T. Hart. 1996. Plant genetic resources collections: an opportunity for the evolution of global data sets. *In* Proc. 3rd Int. Conf./Workshop Integrating GIS and Environ. Model., Santa Fe, NM. 21–26 January. Natl. Center Geogr. Inform. Anal., Santa Barbara, CA (http://www.ncgia.ucsb.edu/conf/SANTA_FE_CD-ROM/main.html).

Greene, S.L., and T.C. Hart. 1999. Implementing geographic analysis in germplasm conservation. p. 25–38. *In* S.L. Greene and L. Guarino (ed.) Linking genetic resources and geography: Emerging strategies for conserving and using crop biodiversity. CSSA Spec. Publ. 27. ASA and CSSA, Madison, WI.

Greene, S.L., T. Hart, and A. Afonin. 1999a. Using geographic information to acquire wild crop germplasm for *ex situ* collections: I. Map development and field use. Crop Sci. 39:836–849.

Guarino, L. 1995. Geographic information systems and remote sensing for the plant germplasm collector. p. 315–328. *In* L. Guarino et al. (ed.) Collecting plant genetic diversity. CAB Int., Wallingford, UK.

Gutterman, Y., and E. Nevo. 1994. Temperature and ecological-genetic differentiation affecting the germination of *Hordeum spontaneum* caryopses harvested from three populations: the Negev and opposing slopes on Mediterranean Mt Carmel. Isr. J. Plant Sci. 42:183–195.

Hamon, S., M. Noirot, and F. Anthony. 1995. Developing a coffee core collection using the principal component score strategy with quantitative data. p. 117–126. *In* T. Hodgkin et al. (ed.) Core collections of plant genetic resources. John Wiley & Sons, Chichester, UK.

Harper, J.L. 1977. The population biology of plants. Acad. Press, London, UK.

Hart, T., S.L. Greene, and A. Afonin. 1996. Mapping for germplasm collections: Site selection and attribution. *In* Proc. 3rd Int. Conf./Workshop on Integrating GIS and Environ. Model., Santa Fe, NM. 21–26 January. Natl. Center for Geogr. Inform. and Anal., Santa Barbara, CA (http://www.ncgia.ucsb.edu/conf/SANTA_FE_CD-ROM/main.html).

Hart, T.C. 1999. Scale considerations in mapping for germ plasm acquisition and the assessment of ex situ collections. p. 51–61. *In* S.L. Greene and L. Guarino (ed.) Linking genetic resources and geography: Emerging strategies for conserving and using crop biodiversity. CSSA Spec. Publ. 27. ASA and CSSA, Madison, WI.

Hawkes, J.G. N. Maxted, and D. Zohary. 1997. Reserve design. p. 132–143. *In* N. Maxted et al. (ed.) Plant genetic conservation: The *in situ* approach. Chapman & Hall, London, UK.

Hong, T.D., and R.H. Ellis. 1996. A protocol to determine seed storage behaviour. IPGRI Tech. Bull. no. 1. IPGRI, Rome, Italy.

Howard, P.C. 1996. Guidelines for the selection of forest nature reserves, with special reference to Uganda. p. 245–262. *In* L.A. Bennun et al. (ed.) Conservation of biodiversity in Africa. Local initiatives and institutional roles. Centre for Biodiversity, Nairobi, Kenya.

Humphries, C. J., P.H. Williams, and R.I. Vane-Wright. 1995. Measuring biodiversity value for conservation. Annu. Rev. Ecol. System. 26: 93–111.

Hutchinson, M.F., H.A. Nix, J.P. McMahon, and K.D. Ord. 1996. The development of a topographic and climate database for Africa. *In* Proc. 3rd Int. Conf./Workshop on Integrating GIS and Environ. Model., Santa Fe, NM. 21–26 January. Natl. Center for Geogr. Inform. Anal., Santa Barbara, CA (http://www.ncgia.ucsb.edu/conf/SANTA_FE_CD-ROM/main.html).

Jones, P.G., S.E. Beebe, J. Tohme, and N.W. Galwey. 1997. The use of geographical information systems in biodiversity exploration and conservation. Biodivers. Conserv. 6:947–958.

Justice, C.O., B.N. Holben and M.D. Gwynne. 1987. Monitoring East African vegetation using AVHRR data. Int. J. Remote Sens. 7:1453–1474.

Kessell, S.R. 1990. An Australian geographic information and modelling system for natural area management. Int. J. Geogr. Inform. Syst. 4:333–362.

Kirkpatrick, J.B. 1974. The use of differential systematics in geographic research. Area 6:52–53.

Legg, C. 1995. A geographic information system for planning and managing the conservation of tropical forests in the Knuckles Range. Sri Lanka For. Spec. Issue:37–62.

Liu, J.G., J.B. Dunning, and H.R. Pulliam. 1995. Potential effects of a forest management plan on Bachman's sparrows (*Aimophila aestivalis*): Linking a spatially explicit model with GIS. Conserv. Biol. 9:62–75.

Maxted, N., M.W. van Slageren, and J.R. Rihan. 1995. Ecogeographic surveys. p. 255–285. *In* L. Guarino et al. (ed.) Collecting plant genetic diversity. CAB Int., Wallingford, UK.

Maxted, N., B.V. Forde-Lloyd, and J.G. Hawkes. 1997a. Complementary conservation strategies. p. 15–39. *In* N. Maxted et al. (eds.) Plant genetic conservation: The *in situ* approach. Chapman & Hall, London, England.

Maxted, N., L. Guarino, and M.E. Dulloo. 1997b. Management and monitoring. p. 144–159. *In* N. Maxted et al. (ed.) Plant genetic conservation: The *in situ* approach. Chapman & Hall, London, UK.

Maxted, N., J.G. Hawkes, L. Guarino, and M. Sawkins. 1997c. The selection of plant conservation targets. Genetic Resour. Crop Evol. 7:1–12.

Menges, E.S. 1991. The application of minimum viable population theory to plants. p. 45–61. *In* D.A. Falk and K.E. Holsinger (ed.) Genetics and conservation of rare plants. Oxford Univ. Press, Oxford, UK.

Miller, R.I. 1986. Predicting rare plant distribution patterns in the southern Appalachians of the southeastern USA. J. Biogeography 13:293–311.

Monestiez, P., M. Goulard, and G. Charmet. 1994. Geostatistics for spatial genetic structures: Study of wild populations of perennial ryegrass. Theoret. Appl. Genet. 88:33–41.

Monmonier, M. 1973. Maximum-difference barriers: An alternative numerical regionalization method. Geogr. Anal. 3:245–261.

Nabhan, G.P. 1991. Wild *Phaseolus* ecogeography in the Sierra Madre Occidental, Mexico. System. Ecogeogr. Stud. Crop Gene Pools no. 5. IBPGR, Rome, Italy.

Perry, M.C., and E. Bettencourt. 1995. Sources of information on existing germplasm collections. p. 121–129. *In* L. Guarino et al. (ed.) Collecting plant genetic diversity. CAB Int., Wallingford, UK.

Pickersgill, B. 1984. Migrations of chili peppers, *Capsicum* spp., in the Americas. p. 105–123. *In* D. Stone (ed.) Pre-Columbian plant migration. Pap. Peabody Museum Archaeol. Ethnol. Vol. 76. Harvard Univ. Press, Boston, MA.

Pigliucci, M., and G. Barbujani. 1991. Geographical pattern of gene frequencies in Italian populations of *Ornithogalum montanum* (Liliaceae). Genet. Res. 58:95–104.

Pollak, L.M., and J.D. Corbett. 1993. Using GIS datasets to classify maize-growing regions in Mexico and central America. Agron. J. 85:1133–1138.

Pollak, L.M., and H.N. Pham. 1989. Classification of maize testing locations in sub-Saharan Africa by using agroclimatic data. Maydica 34:1133–1138.

Rebelo, A.G., and W.R. Sigfried. 1992. Where should nature reserves be located in the Cape Floristic Region, South Africa? Models for the spatial configuration of a reserve network aimed at maximizing the protection of diversity. Conserv. Biol. 6:243–252.

Sackville Hamilton, N.R., and K.H. Chorlton. 1997. Regeneration of accessions in seed collections: a decision guide. Handb. Genebanks no. 5. IPGRI, Rome, Italy.

Salick, J., N. Cellinese, and S. Knapp. 1997. Indigenous diversity of cassava: Generation, use and loss among the Amuesha, Peruvian Upper Amazon. Econ. Bot. 51:6–19.

Scott, J.M., F. Davis, B. Csuti, R. Noss, B. Butterfield, C. Groves, H. Anderson, S. Caicco, F. Dérchia, T.C. Edwards, J. Ulliman, and R.G. Wright. 1993. Gap analysis: A geographic approach to protection of biological diversity. Wildlife Monogr. 123:1–41.

Silvertown, J.W., and J. Lovett Doust. 1995. Introduction to plant population biology. Blackwell Sci. Publ., Oxford, UK.

Skole, D., and C. Tucker. 1993. Tropical deforestation and habitat fragmentation in the Amazon: Satellite data from 1978 to 1988. Science (Washington, DC) 260:1905–1910.

Steiner, J.J., and S.L. Greene. 1996. Proposed ecological descriptors and their utility for plant germplasm collections. Crop Sci. 36:439–451.

Stockwell, D.R.B., and I.R. Noble. 1992. Induction of sets of rules from animal distribution data: A robust and informative method of data analysis. Math. Comput. Simul. 33:385–390.

Tohme, J., P. Jones, S. Beebe, and M. Iwanaga. 1995. The combined use of agroecological and characterization data to establish the CIAT Phaseolus vulgaris core collection. p. 95–107. *In* T. Hodgkin et al. (ed.) Core collections of plant genetic resources. John Wiley & Sons, Chichester, UK.

Vane-Wright, R.I., C.J. Humphries and P.H. Williams. 1991. What to protect? Systematics and the agony of choice. Biol. Conserv. 55:235–254.

Veitch, N., N.R. Web, and B.K. Wyatt. 1995. The application of geographic information systems and remotely sensed data to the conservation of heathland fragments. Biol. Conserv. 72:91–97.

Walker, P.A. 1990. Modelling wildlife distributions using a geographic information system: Kangaroos in relation to climate. J. Biogeogr. 17:279–289.

Waser, N.M. 1987. Spatial genetic heterogeneity in a population of the montane perennial plant *Delphinium nelsonii*. Heredity 58:249–256.

# 2 Implementing Geographic Analysis in Germplasm Conservation

**Stephanie L. Greene**

*Washington State University*
*Prosser, Washington*

**Thomas C. Hart**

*Spatial Data Associates*
*Trumensburg, New York*

Diversity is an intrinsic component of biological systems. The science of conservation biology recognizes the importance of conserving diversity at all levels of biological organization (i.e., ecosystem, species, genes). Plant genetic resources specialists focus primarily on the conservation of genetic diversity (e.g., diversity found within species). Biologists conserving diversity at all levels face the challenge of efficient sampling. Recognizing it is impossible to study every population of every species, conservationists often rely upon indirect information to understand levels and patterns of biodiversity in the landscape. Environmental data are frequently used as surrogate information, under the assumption that habitat variation can be predictive of species diversity (Faith & Walker, 1996). The conservation literature is rich with models that make use of ecogeographic information and spatial analysis to construct "sampling" frames to guide natural reserve establishment (e.g., Lombard et al., 1997; Smith et al., 1997). The significance of environmental variation in influencing intraspecific diversity also is well recognized and remains the "cardinal principle" for sampling genetic diversity (Frankel at al., 1995). The availability of large, digital environmental data sets, and of Geographic Information Systems (GIS) to manipulate them, provides genetic resource professionals with a tool to understand patterns of genetic diversity. However, this must be underpinned by an understanding of the theoretical relationship between geography and the genetic divergence of plant populations.

The fields of geography, ecology and genetics are broad, and have disparate terminologies. It is important to define some concepts and terms to clarify the subsequent discussion. The term "geographic analysis" (GA) refers to the analysis of map-based, remotely sensed or secondarily derived spatial information. Geographic Information Systems may be used to carry out geographic analysis. Guarino et al.

(1999, see Chapter 1) introduces the general concepts of geographic analysis and use of GIS. Other chapters in this volume discuss the use of GA to predict where species naturally occur, or may be successfully introduced. This chapter discusses how population genetics theory can provide guidelines for the effective use of geographic information in sampling genetic resources. The discussion applies to crop landraces, wild and naturalized species, and wild and weedy relatives of crop species. We would expect these classes of germplasm to show genetic differentiation that can be associated to selection pressure from the environment and/or from humans. With respect to selection and gene flow, we include forces attributable to nature and human action. In this chapter, we focus on the unfixed traits contributing to that portion of population genetic structure that reflects adaptation to environmental selection pressures (e.g., the "niche" variation hypothesis tested by Prentice et al., 1995) or that can be attributed to spatial associations such as neighborhood effects (see studies reviewed by Heywood, 1991). The objectives of this chapter are to: (i) review the genetic theory on intraspecific geographic differentiation within the parameters discussed above, and (ii) discuss how projects can be structured to reveal the link between geography and intraspecific genetic diversity.


## FACTORS AFFECTING THE GEOGRAPHIC DIFFERENTIATION OF PLANT POPULATIONS

A plant species can be viewed as a metapopulation, consisting of variously linked populations and subpopulations. Populations and subpopulations are composed of individuals that are grouped close enough to share a similar selective regime and freely intermate (Spieth, 1974). Geographic groupings of individuals frequently originate from a common ancestor (Frankel et al., 1995). A strong body of evidence supports the association between hierarchical levels of population genetic structure and hierarchical levels of geographic structure (Avise, 1994). However, population genetic structure is influenced by a number of microevolutionary forces that are not necessarily associated with the environment per se (Hedrick, 1986).

Genetic divergence between populations (and between subpopulations) is a product of complex evolutionary interactions. The amount of divergence is a function of the relative strength of individual forces. The main evolutionary forces responsible for partitioning genetic variation among and within populations are selection, gene flow and genetic drift; genetic variation is ultimately provided by mutation (Crow, 1986; Hartl, 1988). The type of mating system and other life history traits of the species also have a profound impact on the genetic structure of populations, in interaction with these forces (Loveless & Hamrick, 1984).

### The Influence of Selection

Selection, both natural and artificial, has been recognized since Darwin's time, as a primary driving force in population differentiation. Darwin and Wallace (Ennos, 1990) distinguished two types of selection: abiotic selection (physical selection), caused by soil, climate, etc., and biotic selection, caused by interaction with other plants (competition), pollinators, pathogens and herbivores. Abiotic selection pressure has been associated with the genetic differentiation of plant populations in numerous studies. Excluding the influence of gene flow, drift, and mating systems, the degree and rate of geographic differentiation in a heterogeneous environment increases directly with the intensity of local selection, and with the magnitude of difference between local intensity and regional intensity (defined as the average selection intensity across environmental patches weighted by the relative abundance of given patches; Spieth 1979). Various studies have supported these theories. For example, Snaydon and Davies (1982) observed genetic divergence among populations of *Anthoxanthum odoratum* grown in plots amended with different levels of fertilizer and lime. Bradshaw (1975) summarized a number of studies that demonstrated how strongly opposing edaphic selection regimes can influence differentiation, even at a microgeographic scale. In a classic study large differences in metal tolerance in subpopulations of *Anthoxanthum odoratum* growing only 3 m apart on land bordering a zinc mine was demonstrated (Antonovics & Bradshaw 1970).

Environmental selection pressure also can be associated with genetic differentiation on a broader scale. Steiner and Poklemba (1994) classified 79% of 28 *Lotus corniculatus* accessions into two groups that corresponded with Bailey's highland and lowland ecoregion classification. Bailey's Ecoregion Classification of the Continents is based on climate, soils, landform and potential vegetation (Bailey, 1996). Highland and lowland classes of *L. corniculatus* were differentiated based on differences in high salt-soluble globulin peptides. Pederson et al. (1996) positively correlated the production of hydrocyanic acid in the U.S. white clover collection to climate, using Leemans and Cramer's (1992) global climate data set.

Biotic selection also can strongly influence the genetic structure of populations. Burdon (1980) reported microscale differentiation of a single population of *Trifolium repens* growing in a pasture with homogeneous soil conditions. Differentiation was attributed to the biotic selection pressure exerted by neighboring plants. Although biotic selection has a spatial component, it will usually be difficult to detect on a geographic scale. However, there are biotic pressures that can be associated with the current resolution of geographic information. These include selection caused by plant pathogens associated with a specific area or environmental regime, and anthropogenic selection pressures associated with a specific community or culture.

## The Homogenizing Force of Gene Flow

Gene flow refers to the movement of genes from one population to another. This generally acts to counter the effect of differentiation (Slatkin, 1987). There are many mechanisms responsible for gene movement: pollen dispersal, seed dispersal, migration, even the movement of extranuclear segments of DNA (Slatkin, 1985). In agronomic crops, gene flow is evident between cultivated forms and in the introgression of genes between wild and cultivated forms. Gene flow can occur between species as well. Considering only the evolutionary forces of selection and gene flow, the degree of population divergence depends on the balance achieved by the opposing forces of selection and gene flow (Fig. 2–1). Generally, if selection pressure is very intense, or populations are geographically isolated, gene flow will have little effect on population differentiation (Hamrick, 1987). In spite of gene flow, allele frequencies will revert back to original levels under strong selection pressure and will tend to become fixed in the complete absence of gene flow. Gene flow can be a source of variability according to Wright's shifting balance theory (Slatkin, 1985; Hamrick, 1987). Wright proposed that gene combinations that enhance fitness can become partially or completely fixed in a population when selection and genetic drift overcome low levels of gene flow or when gene flow is absent. Demographic changes that re-establish or increase gene flow can then result in the movement of adaptive gene combinations into adjacent populations. In a heterogeneous environment, where selection pressures are similar and moderate to neutral, a number of models have demonstrated that as rate of gene flow increases, geographic differentiation decreases. At a critical point, the polymorphism can become unstable and disappear (Felsentein, 1976). When selection pressures are weak, even minimal gene flow between two populations can prevent population divergence (Hamrick, 1987).

Because plants are immobile, and because of the difficulty in quantifying the effects of long-distance gene flow (Hamrick, 1987), the influence of gene flow on the evolution of populations was once thought to be relatively minor (e.g., Griffiths, 1950; Bradshaw, 1977). Even rare, long-distance dispersal can strongly limit population differentiation (Hamrick, 1987). Ecogeographic information may be helpful in understanding the potential for long-distance gene flow between populations. For example, two plant populations far enough apart to prevent pollen exchange by insects, but growing along a dispersal route such as a road, trail or stream, are more likely to experience gene flow due to long distance seed and/or pollen dispersal by animals, humans or water. Conversely, if two populations are not on the same dispersal route, for instance occurring in different drainage systems, long-distance gene flow would be less likely.

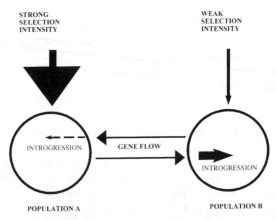

Fig. 2–1. Simple model of intraspecific genetic differentiation between two populations with the same rate of gene flow, but exposed to strong and weak selection pressure. Population A will have little introgression with Population B since strong selection pressure will return allele frequencies to levels of the adapted Population A. Population B will have greater introgression with Population A since the selection intensity is not strong enough to overcome the influence of gene flow (given that alleles from Population A convey the same level of fitness as alleles from Population B).

## Genetic Drift, Population Bottlenecks, and Founder Events

Gene flow can readily overcome the effects of background or stochastic genetic change, known as genetic drift. The extent to which drift will result in population divergence is influenced by effective population size ($N_e$), the number of individuals that can contribute a gamete to the next generation. As effective population size decreases, the greater the impact drift has on population differentiation. Gene flow overcomes the effect of drift since it increases $N_e$. A rule of thumb put forth by Spieth (1974) based on Wright's classic work, is that one migrant individual per local population, per generation, will prevent population divergence due to drift. Demographic changes such as a reduction in population size resulting in a population bottleneck, or the colonization of a new area by a small number of individuals (founder event), have the same effect as random drift in changing allele frequencies: there will be less divergence in a newly restricted or colonized population if $N_e$ remains relatively large through gene flow.

## Life History Traits that Influence Geographic Differentiation

Life history traits influence the extent that populations within a species can become geographically distinct by influencing gene flow and effective population size. Loveless and Hamrick (1984) reviewed the literature to determine the relative influence that ecological factors have on population genetic structure. After reviewing data of mating systems ranging from obligate autogamy to obligate allogamy, they found that inbreeding systems tended to promote population differentiation, while outbreeding systems tended not to promote differentiation between populations. Floral morphology, mode of reproduction and pollination mechanism all influenced the extent of population differentiation. For example, species pollinated by insects showed greater population differentiation on average, due to relatively limited pollen dispersal, than did species pollinated by larger animals or by wind, where pollen was dispersed across a wider area. Seed dispersal and seed dormancy had a lesser but still significant influence on population differentiation. Species with seeds dispersed by gravity tended to have greater population differentiation than species with seeds dispersed by animal, wind or water. Although these ecological variables generally influenced gene flow, some, such as seed dormancy, influenced effective population size by increasing the soil seed bank.

## The Assumption of Genetic Equilibrium

An important assumption of these genetic theories is that geographic differentiation reflects equilibrium allele frequencies (Spieth, 1979; Heywood, 1991). In order to use such theories to associate patterns in a conclusive manner, researchers need to work with populations that are in genetic equilibrium. This entails understanding landscape disturbances and defining a measure of "recency" based on the length of time it might take the target populations to reach equilibrium. The length of time it takes for equilibrium to be established within a population depends upon the type and intensity of evolutionary forces that initiate the change. Life history traits such as generation time and mating system also will influence the rate at which

Table 2–1. Phases for implementing a project to sample germplasm using geographic information and GIS analysis.

| Phases | Focus |
|--------|-------|
| I.   Project initiation | Based on project objectives, is there value in investing in a geographic analysis? |
| II.  Database development | Chou (1996); Hart (1999, see Chapter 4) |
| III. Database analysis | Where should we sample? |
| IV.  On the ground | Describe site at local level |
| V.   Post-trip database analysis | Refine database. What did we sample? |

equilibrium is established. Computer simulations indicate that local population structure could evolve in 10 to 20 generations following a disruptive event (Heywood, 1991). Recently disturbed areas such as roadsides, logged or newly grazed areas or natural landslide or flood areas are likely to support populations that do not yet reflect current regimes of selection and gene flow (Heywood, 1991).

## Successfully Linking Geography and Genetic Resources

A geographic analysis focused on developing a framework for sampling germplasm for ex situ conservation can be divided into five phases (Table 2–1). In each phase, researchers need to consider how genetic theories of geographic differentiation can be applied. We will discuss the successful and unsuccessful application of theory as it relates to project initiation, pre- and postcollection GIS database query, and collection in the field. The use of input data and database development are critical for project success. For a more general treatment of database development see Chou (1996).

## Project Initiation

The objective of the initial project phase is to determine that a satisfactory return can be expected from the investment needed to implement a geographic analysis. Effective application demands that GIS projects be driven by conservation needs, and not the need to make use of impressive new tools. The first step is to develop clear objectives that reflect conservation priorities. Researchers then need to assess the specifics of the targeted exploration area, the taxa to be collected, and sampling objectives, to determine how these factors will affect the use of geographic information as surrogate indications of genetic diversity. Table 2–2 provides a list of key questions that should be asked during the initial assessment. If project objectives are focused on sampling patterns of diversity that are difficult or impossible to link to geography, then the extra effort and expense of a geographic analysis will probably not be warranted.

Geographic information will be more effective when targeted genetic diversity is influenced by evolutionary forces that promote the geographic differentiation of plant populations. Theoretically, this is most likely to occur in heterogeneous landscapes that provide a mosaic of selection regimes. Generally, the stronger the difference between regimes, the greater the resident populations would diverge. An example of a landscape characterized by sharp ecogeographic differences between

Table 2–2. Factors to consider in determining the potential value of using geographic information to understand patterns of intraspecific genetic diversity.

| Factor | |
| --- | --- |
| Targeted geographic area | |
|     Geography | Size of area |
| | Landscape heterogeneity |
| | Magnitude of difference between landscape patches |
| | Intensity of abiotic selection |
|     Ecology | Agricultural or natural ecosystem |
| | Degree of stability |
| | Recency of disturbance (natural or anthropogenic) |
| | Level of biodiversity |
| | Value as a conservation unit |
| | Intensity of biotic selection |
|     Data coverage | Quality |
| | Quantity |
| | Availability |
| Targeted taxon | |
|     Life history traits | Breeding system |
| | Method of dispersal |
| | Longevity, phenology |
| | Habitat specificity |
|     Distribution | Narrow or widespread |
| | Rare or abundant |
| | Contiguous or patchy |
|     Germplasm category | Improvement status (e.g., landrace, wild or weedy relative) |
| | Agronomic status (e.g., wild, cultivated, native or introduced) |
| | Level of vulnerability (e.g., threatened, endangered, common) |
|     Sampling objective | Strategy (in situ, ex situ) |
| | Targeted trait (e.g., disease, insect resistance, adaptation, quality) |
| | Control of trait (e.g., polygenic or monogenic) |

patches would be southeastern Utah. This is a high desert characterized by a mosaic of different moisture, soil and biotic regimes, the results of erosion fragmenting the Colorado Plateau into numerous young canyon systems. Mountainous terrain, complex in its patterns of elevation, aspect and slope, also would have patches where strong local abiotic pressure could overcome the leveling effects of gene flow.

Most landscapes are not as obvious. An understanding of the geology and natural vegetation can be helpful to collectors. For example, Fig. 2–2 illustrates the relatively homogeneous environment of the Horse Heaven Hills that extend along the southern boundary of the Columbia Plateau in Washington. The hills consist of deep Palouse loess soils and have an open, continuous grassland vegetation. Given the relatively uniform selection regime we would expect little geographic differentiation to have occurred among plant populations, especially for wind-pollinated grasses, where gene flow would be strong. In contrast, Fig. 2–3 illustrates an area 150 km to the northeast. Although still part of the Columbia Plateau, this area is characterized by islands of Palouse loess separated by basalt coulees, formed when great floods scoured the area, forming the channeled scablands. In addition, this area occurs in the transition zone between grasslands and conifer forests. This type of fragmented landscape would be more likely to give rise to geographic differentiation between plant populations growing throughout the area.

Fig. 2–2. Example of a homogeneous landscape with uniform soil and vegetation in the Horse Heaven Hills, Washington. Geographic differences would be less likely to occur among populations exposed to similar selection regimes.

Plant populations isolated by either distance or intervening land forms could be expected to diverge even if selection regimes did not contrast sharply. Researchers need to consider how the occurrence of dispersal routes (roads, trails, rivers, landscape corridors) and geographic distance could influence gene flow. In

Fig. 2–3. Example of heterogeneous landscape likely to support the geographic differentiation of plant populations. Islands of Palouse loess bisected by the channeled scablands, eastern Washington.

the case of landraces where introgression is occurring between wild and cultivated forms, where similarities exist in crop use between communities, or where road systems promote movement of germplasm, geographic information may not accurately reflect patterns of intraspecific diversity. Conversely, in cases where the target region houses fragmented and geographically or culturally isolated communities, whose landraces are subject to effects of selection and drift that can be georeferenced, then patterns of geographic diversity may well effectively reflect patterns of genetic diversity.

In determining whether a geographic approach is worthwhile, the breeding system and life history traits of target species are also important considerations. Although self-pollinated species are more likely to reflect geographic differentiation, population divergence has been reported in cross-pollinated and wind-pollinated species and attributed to environmental adaptation (Prentice et al., 1995; Rumbaugh et al., 1988). Life history traits help define the resolution of the expected genetic pattern. For example, if the objective is to sample microgeographic differentiation, a geographic analysis would be more effective for self-pollinated target species than for wind-pollinated, given the same level of selection. These considerations are important in determining the resolution that will be needed for the geographic information.

From a practical standpoint, the availability and quality of map information is an important consideration in evaluating the development of a GIS database. For example, in collecting wild forage and vegetable germplasm from Albania, the authors decided that a GIS database would not be worth the expense since the most recent satellite imagery of the targeted area dated from 1991 and it was expected that other thematic coverages would be even older. Considering the pace of recent socioeconomic and political changes in the country and degree of genetic erosion reported by previous explorers (Hammer et al., 1996), we decided the data were not timely enough to accurately identify areas that had relatively little anthropogenic disturbance. Aerial photos that provided information on the location of cemeteries, wood lots and other areas where grazing would be controlled would have been ideal, but were simply not available for this country.

Another practical consideration in determining the potential value of carrying out a geographic analysis is the nature of the targeted trait. If expression of the specific trait is independent of the environment, there is little reason to develop a GIS database, except as a navigational tool for collectors. For example, geographic information may not provide a useful framework to sample germplasm with a specific flower color, cultural use or pest resistance.

In areas that have experienced recent disturbance or change (fire, logging, urbanization), researchers need to judge how quickly the target species could be expected to reach genetic equilibrium against the length of time since the disturbance. For example, in 1996 we collected the endangered species *Trifolium thompsonii* in central Washington. In 1988, a fire had swept through all existing populations but one, and these populations have since been spreading (L. Julian, 1996, personal communication). In this instance, a geographic sampling framework would probably not correspond with patterns of genetic diversity since genetic equilibrium most likely has not been reached, considering the magnitude of ecosystem change, and limited number of generations expected from this slow growing perennial, since the fire.

A final practical consideration in determining the value of developing a GIS database is the potential use of the database by other scientists. Geographic Information Systems databases developed in areas that are known centers of crop diversity or otherwise rich in biological diversity are likely to be used by others, which would increase the expected returns of investing in a geographic analysis. For example, a detailed GIS database was developed for the floristically rich Western Caucasus Mountains in Russia (Greene et al., 1999a). Although the total cost of the project was $30 000, the GIS information was used to guide the field collection of wild forage species and to provide extensive passport data for the accessions collected. It has subsequently been used in three research projects that have studied the collected germplasm. The data also were used to determine the appropriate timing to evaluate the alleged cold-hardiness of the Caucasus honeybee (Lundeen, 1997, personal communication). Additionally, there is interest in using the data to collect such diverse germplasm as apple (*Malus domestica* Borkh; Forsline, 1996, personal communication) and onion (*Allium cepa* L. var. *cepa*; Afonin, 1997, personal communication). Since this GIS database has been placed on a CD-ROM, it is readily available.

## Identification of Potential Sites

Once the database has been developed, it can be used to identify areas where plant populations may be geographically differentiated. The specific precollection assessment will depend on the geographic data that has been assembled. For example, if high-quality soil classification maps are available, soil boundaries could be examined. Theoretically, populations growing adjacent to boundaries would reflect less edaphic adaptation than populations growing within the center of patches as a result of gene flow and/or constantly changing selection regimes from a variable environment (Spieth, 1979). Populations growing in small patches, but surrounded by an edaphic regime that favored different alleles, would theoretically be able to maintain the adaptive allele in only low frequencies, if at all (Felsenstein, 1976). Decisions regarding where to sample populations can thus be guided by applying genetic theory to geographic information.

A useful application that is now practical is the identification of areas rich in geographic diversity. The researcher can thus increase the efficiency of collecting itineraries and identify areas were in situ reserves could most usefully be established (Greene et al., 1999a). Most GIS software systems support pattern analysis that makes this type of diversity measurement straightforward. Choosing whether to use empirical measurements specific only to the site under study, or whether to use the widely applicable standard measurements of diversity used by landscape ecologists is a difficult decision. More arguments seem to favor the use of empirical measurements. First, the environmental range of a particular study area is often much smaller than the range in which standard measurements were developed, and as such, the standard measures may miss important local variants. Using the Russian Caucasus Mountain project as an example, empirical climatic data classified the region into 15 moisture zones and 14 temperature zones. Using the standard Bailey's

Ecosystem Classification of the Continents, the same area was classified into three ecoregions (Bailey, 1996). For the purpose of collecting diverse germplasm, empirical data were the most useful. Second, the data required to support the development of standard measurements may not be available for the area of interest. Substitute computations using different available information may be required. On the other hand, standard measurements allow for comparisons to be drawn between studies and across entire programs. In the end, it may be advisable to seek both types of results, and as data become more sophisticated and accessible, the standard measures will grow in feasibility and effectiveness.

In this phase of the project, the researcher will start to map hypothetical boundaries defining the priority target areas within the study region. It is critical at this stage to recognize that population boundaries are going to be a function of geographic as well as nongeographic forces such as life history traits and demographic factors. Although it may be difficult or impossible to assess the influence of past demographic changes such as bottlenecks or founder events, researchers can readily take into account the life history traits of the target germplasm in estimating boundaries. For example, if a collector was sampling a wind-pollinated grass in a homogeneous landscape, a single interbreeding population might consist of all individuals over a wide area. If the target species was clonally propagated or reproduced apomictically, multiple populations could be defined among groups of individuals across the same area.

## Support of Fieldwork

During the course of field visits to collecting sites, researchers need to pay close attention to two factors that will influence the success of using the GIS database to understand patterns of intraspecific diversity. The first is whether populations are in genetic equilibrium with the environment. Local people could be questioned regarding the age of existing populations, the frequency and intensity of disturbances and the local history of land use. In the case of wild species, associated vegetation can be assessed for the presence of weedy or invasive species that might suggest the area has been recently disturbed (Greene et al., 1999b). Evidence of overgrazing, logging, road building, recent landslides, etc., could suggest recent disturbance of the area. Collecting from roadsides is frequently very productive, easily accessed, and often rich in indigenous plant species. Collectors need to bear in mind, however, that accessions collected from these areas may be recent migrants and thus may not reflect adaptation to the site, in terms of having a high frequency of locally common alleles (Greene et al., 1999b). Roadside verges also are likely to show microclimatic differences compared to the surrounding habitat, which leads to the second consideration. If accessions are collected from small patches of isolated microhabitat, genetic patterns will not necessarily be reflected by the geographic database. Hart (Hart, 1999, see Chapter 4) describes the occurrence of "inlier" sites in more detail. If samples are taken from such sites, researchers need to recognize the danger of using geographic information at broad resolution to describe the site. If material is collected there, local documentation will be the only source of information describing the site at the appropriate resolution. In this instance, detailed information describing microsite deviation needs to be recorded using forms such as the one proposed by Steiner and Greene (1996).

## Post-Trip Analysis

Information obtained during fieldwork can be used to further refine the GIS database. For example, Hart et al. (1996) found that soil and road maps were not detailed or recent enough to reflect collecting opportunities in the Caucasus Mountains of southern Russia. These data sets were therefore not used further. The remaining GIS data were then used to comprehensively describe each of the collected sites, in addition to site-specific soil data collected in the field (Hart et al., 1996). The refined database can be used to quantify collection coverage, to identify eco-geographic redundancy in sites and accessions sampled, and to identify accessions that could be uniquely adapted to specific environments. This type of analysis may be useful for identifying accessions that contribute new and useful variation to a collection, in cases where many samples have been collected of a given taxa but collection managers cannot justify the inclusion of all samples into an ex situ collection due to resource constraints (Greene et al., 1999b). By applying population genetic theory to geographic and site-specific information, curators and germplasm users have a cost-effective procedure for understanding the germplasm that has been collected. The results are improved decision-making based on the rational consideration of which accessions will contribute the most to an ex situ collection, and ultimately to the evaluation and further use of material in breeding programs.

The refined database also can be used to suggest where to locate in situ reserves. Geographic information can help identify areas that are particularly heterogeneous, or identify populations that are adapted to unique environments or that may be vulnerable to extinction. Additional information can be added to the database to facilitate the decision-making process. For example, a layer depicting land ownership may be useful at this point. Maxted et al. (1997) outline the development of in situ reserves. There are many points during this process where geographic information, spatial analysis and concepts of population genetics can be integrated to inform the decision-making process.

## CONCLUSIONS

We have presented a brief summary of population genetic theory to provide guidelines for using geographic patterns to efficiently sample patterns of intraspecific genetic diversity. Sampling environmental variability is a key strategy for preserving plant genetic diversity (Frankel et al., 1995). The effective sampling of environmental adaptation also is important if breeders are to succeed in developing crop varieties adapted to marginal conditions. The mainstream application of spatial analysis in decision-making processes has only developed within the last 15 yr. The application of geographic information and GIS analysis to the conservation of plant genetic resources is only now moving from conception to application. At this point, research is needed to develop methods demonstrating the relationship between geography and population differentiation. These methods need to be used to assess the strength and frequency of such relationships and determine if germplasm sampling driven by GIS-derived information is more effective than other approaches, such as random sampling. This research will be motivated by the increased need for efficient and effective conservation, coupled with the inevitable increase in available geographic information and ease of carrying out geographic analysis.

## REFERENCES

Antonovics, J., and A.D. Bradshaw. 1970. Evolution in closely adjacent plant populations. VII. Clinal patterns at a mine boundary. Heredity 24:349–362.

Avise, J.C. 1994. Molecular markers, natural history and evolution. p. 204–205. Chapman and Hall, New York.

Bailey, R.G. 1996. Ecosystem geography. Springer-Verlag, New York.

Bradshaw, A.H. 1975. Population structure and the effects of isolation and selection. p. 37–51. In O.H. Frankel and J.G. Hawkes (ed.) Crop genetic resources for today and tomorrow. Cambridge Univ. Press, Cambridge, UK.

Burdon, J.J. 1980. Intra-specific diversity in a natural population of Trifolium repens. J. Ecol. 68:717–735.

Crow, J.F. 1986. Basic concepts in population, quantitative and evolutionary genetics. W.H. Freeman and Company, New York.

Ennos, R.A. 1990. Detection and measurement of selection: genetic and ecological approaches. p. 200–214. In A.H.D. Brown et al. (ed.) Plant population genetics, breeding and genetic resources. Sinauer Assoc., Inc., Sunderland, MD

Faith D.P., and P.A. Walker. 1996. Environmental diversity: On the best-possible use of surrogate data for assessing the relative biodiversity of sets of areas. Biodivers. Conserv. 5:399–415.

Frankel, O.H., A.H.D. Brown, and J.J. Burdon. 1995. The conservation of plant biodiversity. Cambridge Univ. Press, Cambridge, UK.

Felsenstein, J. 1976. The theoretical population genetics of variable selection and migration. Annu. Rev. Genet. 10:253–280.

Govindaraju, D.R. 1988. Relationship between dispersal ability and levels of gene flow in plants. Oikos 52:31–35.

Greene, S.L., T.C. Hart, and A. Afonin. 1999a. Using geographic information to acquire wild crop germplasm: I. Map development and field use Crop Sci. 39:836–842.

Greene, S.L., T.C. Hart, and A. Afonin. 1999b. Using geographic information to acquire wild crop germplasm: II. Post collection analysis. Crop Sci. 39:843–849.

Griffiths, D.J. 1950. The liability of seed crops of perennial ryegrass (Lolium perenne) to contamination by wind borne pollen. J. Agric. Sci. 40:277–284.

Guarino, L. 1995. Geographic information systems and remote sensing for plant germplasm collectors. p. 315–328. In L. Guarino et al. (ed.) Collecting plant genetic diversity: technical guidelines. CAB Int., Wallingford, United Kingdom.

Guarino, L., N. Maxted, and M. Sawkins. 1999. Analysis of georeferenced data and the conservation and use of plant genetic resources. p. 1–24. In S.L. Greene and L. Guarino (ed.) Linking genetic resources and geography: Emerging strategies for conserving and using crop biodiversity. CSSA Spec. Publ. 27. ASA and CSSA, and SSSA, Madison, WI.

Hammer, K., G. Laghetti, G. Olita, P. Perrino, and L. Xhuveli. 1996. Collecting in the Albanian mountains, 1995. Plant Genet. Resour. Newsl. 107:36–40.

Hamrick, J.L. 1987. Gene flow and distribution of genetic variation in plant populations. p. 53–67. In M. Urbanska (ed.) Differentiation patterns in higher plants. Acad. Press, New York.

Harlan, J.R., and J.M.J. de Wet. 1971. Toward a rational classification of cultivated plants. Taxon 20:509–517.

Hart, T.C. 1999. Scale considerations in mapping for germplasm acquisition and the assessment of ex situ collections. p. 51–61. In S.L. Greene and L. Guarino (ed.) Linking genetic resources and geography: Emerging strategies for conserving and using crop biodiversity. CSSA Spec. Publ. 27. ASA and CSSA, Madison, WI.

Hart, T.C., S.L. Greene, and A. Afonin. 1996. Mapping for germplasm collections: Site selection and attribution. In Proc. 3rd Int. Conf. Integrat. GIS and Environ. Model., 21–25 January. NCGIA, Santa Fe, New Mexico. (available on-line at http://www.ncgia.ucsb.edu/conf/SANTA FE CD-ROM/main.html).

Hartl, D.L. 1988. A primer of population genetics. 2nd ed. Sinauer Assoc., Inc., Sunderland, MA.

Hedrick, P.W. 1986. Genetic polymorphism in heterogeneous environments: A decade later. Ann. Rev. Ecol. Syst. 17:535–566.

Heywood, J.S. 1991. Spatial analysis of genetic variation in plant populations. Ann. Rev. Ecol. Syst. 22:335–355.

Leemans, R., and W.P. Cramer. 1992. The IIASA database for mean monthly values of temperature, precipitation, and cloudiness on a global terrestrial grid. Digital raster data on a 30 minute geographic (lat/long) 360 X 720 grid. Thirty-six independent single-attribute spatial layers on CD-ROM, 15.6 MB. Version 1.0. Disc. A. Global Ecosyst. Database, Boulder, CO.

Lombard, A.T., R.M. Cowling, R.L. Pressey, and P.J. Mustart. 1997. Reserve selection in a species-rich and fragmented landscape on the Agulhas Plain, South Africa. Conserv. Biol. 11:1101–1115.

Loveless, M.D., and J.L. Hamrick. 1984. Ecological determinants of genetic structure in plant populations. Ann. Rev. Ecol. Syst. 15:65–95.

Maxted, N., B. Ford-Lloyd, and J.G. Hawkes (ed.). 1997. Plant genetic conservation: The *in situ* approach. Chapman and Hill, New York.

Prentice, H.C., M. Lönn, L.P. Lefkovitch, and H.Runyeon. 1995. Associations between allele frequencies in *Festuca ovina* and habitat variation in the alvar grasslands on the Baltic island of Öland. J. Ecol. 83:391–402.

Pederson, G.A., T.E. Fairbrother, and S.L. Greene. 1996. Cyanogenesis and climatic relationships in the U.S. white clover germplasm collection. Crop Sci. 36:427–433.

Rumbaugh, M.D., W.L. Graves, J.L. Caddel, and R.M. Mohammad. 1988. Variability in a collection of alfalfa germplasm from Morocco. Crop Sci. 28:605–609.

Slatkin, M. 1985. Gene flow in natural populations. Ann. Rev. Ecol. Syst. 16:393–430.

Slatkin, M. 1987. Gene flow and the geographic structure of natural populations. Science (Washington, DC) 236:787–792.

Smith, A.P., N. Horning, and D. Moore. 1997. Regional biodiversity planning and Lemur conservation with GIS in Western Madagascar. Conserv. Biol. 11:498–512.

Snaydon, R.W., and T.M. Davis. 1982. Rapid divergence of plant populations in response to recent changes in soil conditions. Evolution 36:289–297.

Spieth, P.T. 1979. Environmental heterogeneity: A problem of contradictory selection pressures, gene flow and local polymorphism. Am. Nat. 113:247–260.

Spieth, P.T. 1974. Gene flow and genetic differentiation. Genetics 78:961–965.

Steiner, J.J., and C.J. Poklemba. 1994. *Lotus corniculatus* classification by seed globulin polypeptides and relationship to accession pedigrees and geographic origin. Crop Sci. 34:255–264.

Steiner, J.J., and S.L. Greene. 1996. Proposed ecological descriptors and their utility for plant germplasm collection. Crop Sci.36:439–451

# 3    Exploring the Relationship of Plant Genotype and Phenotype to Ecogeography

Jeffrey J. Steiner

*USDA-ARS, National Forage Seed Production Research Center*
*Corvallis, Oregon*

The view that knowledge about the genetic relationships of plants to the environments in which they grow can be used for plant improvement is not new and has led to debate regarding how germplasm resources should be collected, maintained, and utilized (Pistorius, 1997). However, there are many ways these kinds of information can be put to use, regardless of whether plant breeding strategies are based on single-gene or polygenetic approaches. Personal field observations led Vavilov (1992) in 1932 to propose using information on the environments as a way to conduct plant germplasm collection expeditions. He believed that this would allow collections to be made in shorter periods of time and more economically. It was Rick's (1973) perspective that native habitat information describing where germplasm naturally occurred could provide insights into plant adaptations that could be useful for crop improvement. A practical use of the understanding of how plants respond in different environments may be the possibility to predict how a species will respond to a set of management conditions (Bailey, 1989).

During the 1990s, significant technical advancements have made molecular biology and geographic information system (GIS) tools readily available for use in germplasm resource management and research. In addition to these developing technologies, progress has been made in updating and upgrading the kinds of data that are available for describing ex situ germplasm collections. By integrating what we already know about germplasm holdings with purposeful exploration using these tools, the germplasm user community has the possibility to know more about each accession than ever before, so that these valuable resources can be more efficiently utilized. It has been suggested that elucidating the relationship between the genetic makeup of accessions in ex situ germplasm collections and descriptions of their natural ecological adaptation may be a way to identify germplasm that contains desired or novel traits, but this area of research has not been extensively explored (Greene & McFerson, 1994; Steiner, 1994; Steiner & Poklemba, 1994; von Bothmer & Seberg, 1995).

The purpose of this chapter is to: (i) provide a brief background on how the genetic composition of germplasm collected in the wild is influenced by the natural environments from which it is collected; (ii) present some methodological approaches to the analysis of the relationship between ecogeography and genotype/phenotype, and (iii) demonstrate the utility of these approaches through a series of examples using temperate forage legume species.

## NATURAL ENVIRONMENTS, PLANT DISTRIBUTIONS AND ECOLOGICAL CLASSIFICATIONS

The presence of different kinds of vegetation in a landscape is the result of an integration of the effects of many different environmental factors (Akin, 1991), with climate a particularly important controlling factor (Walter, 1985). Though species may respond differently to different environmental factors and different environmental factors may limit the range of a species in different regions (Kuchler, 1988), the different kinds of plants that inhabit a landscape are considered to be in equilibrium with the climate (Woodward, 1987). Numerous classification systems have been devised by ecologists to describe the climate-determined assemblages of natural vegetation across broad regions. Examples of these include A Classification of the Biogeographical Provinces of the World (Udvardy, 1975) and Ecoregions of the Continents (Bailey, 1989).

The usefulness of such ecoregional classifications depends greatly upon their intended use (Omernik & Gallant, 1990). For plant germplasm resource workers, the biggest problem with ecoregional classifications comes from local-scale features that are not shown on large-scale maps. Broad geographic areas are really mosaics whose characteristics vary with scale (Omernik, 1994). Secondary factors such as microenvironment can exert strong control over plant distributions (Good, 1964). Even though climate determines the potential of a species to occupy a given location (Akin, 1991), smaller-scale edaphic and topographic features, or disturbances of different kinds, can significantly affect the actual presence or absence of a species in a landscape (Fosberg, 1967). The practical effect of this is that a species may only occur in a limited area within a broader ecoregion, as a result of a geographic inlier or disturbance to the site (e.g., a riparian zone within an arid region, or along a roadway cut through a forest, respectively).

When such features as these are encountered at the time of accession acquisition, they must be noted on collecting forms or the interpretation of germplasm classifications based on agronomic, physiological, or biochemical measures may be difficult to validate when analyzed at a later time (Steiner & Greene, 1996). These kinds of detailed collecting site descriptions have often been lacking in passport data, but can be invaluable to users of ex situ germplasm collections when attempting to explain genotypic and phenotypic relationships among accessions. Landscape-level ecogeographic descriptions have been used to classify germplasm collections (Steiner & Poklemba, 1994) and to demonstrate associations between phylogenetic relatedness and ecological distributions (Steiner, 1999). Using a historic birdsfoot trefoil (*Lotus corniculatus* L. var. *corniculatus*) data set (Chrtkova-Zertova, 1973), it was possible to use cladistic analysis of morphological features not only to discriminate between naturally occurring subspecies by warm and cool habitats, but also to show a phylogenetic gradient of subspecies from cool southern European alpine regions to severe cold environments in northern boreal regions (Fig. 3–1). Analyses such as these give insights into morphological features that may assist plant breeders to identify key traits for selecting appropriate germplasm suited to different environments.

## Using Ecogeographic Data

In summary, ecogeographic information is helpful in understanding the ecological and genetic structures of germplasm collections. However, generalized ecoregional descriptions do not provide exact or approximated information on the range of climatic conditions found within ecoregions. Also, it is not easy to deter-

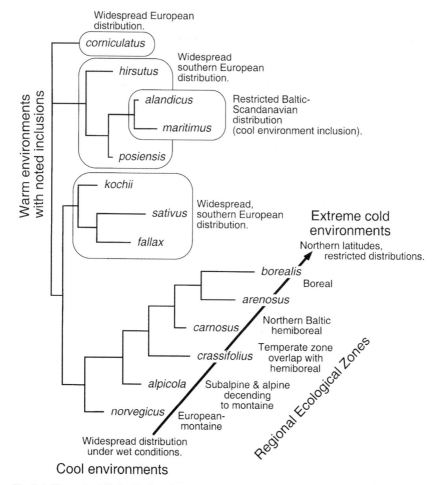

Fig. 3–1. The ecological distribution of central and northern European birdsfoot trefoil subspecies and their phylogenetic relatedness based on morphological characteristics. This figure is adapted from Steiner (1999). The original data set was obtained from Chartkkova-Zertova (1973).

mine which of the environmental features that characterize an ecoregion actually influence the presence or absence of specific genotypes or morphological traits at a collection site. These drawbacks of qualitative ecogeographical information can be overcome by using complementary quantitative climatic and geographic descriptions of collecting sites. Unfortunately, such kinds of environmental information are rarely available from field site documentation.

An approach to these problems is to obtain collecting site data, after-the-fact given their latitude and longitude, by using GIS software such as IDRISI (1993) in conjunction with global-scale data sets of qualitative and quantitative climatic and ecological information such as the U.S. Environmental Protection Agency (EPA) and National Oceanic and Atmospheric Administration (NOAA) Global Ecosystems Databases (Kineman & Ohrenschall, 1992, 1994). When geographic coordinates are not available for the collecting sites, field notes and atlases can be used to estimate locations (Steiner & Greene, 1996).

However, it is important to be realistic about the accuracy and limitations of global-scale databases when making interpretations (Estes & Mooneyhan, 1994). The relative accuracy of the global map data should be assessed by comparison with on- or near-site data sources whenever possible (Steiner & Greene, 1996). Also, relevant data more suitable to regional or other geographic scales may be obtained from regional sources (e.g., USDA Natural Resour. Conserv. Serv.) for more accurate estimates than would be possible using global-scale sources. It also is possible to build a custom data set for a particular collection region using other geographic tools and resources (Hart et al., 1996).

## Quantitative Approaches to Exploring Plant Relationships with Ecogeographic Data

Foresters have been particularly active in investigating the relationship between plant distributions and ecogeographic data. Quantitative relationships between physical environmental factors and plant distributions have been described and the validity of region-level plant distribution descriptions tested using quantitative approaches such as discriminant analysis (Zobler, 1957) and cluster analysis (Radloff & Betters, 1978). Quantitative approaches have been extended to predict the productivity of forest ecosystems based on pre-established boundaries determined independently of productivity estimates (Baily, 1984) and to estimate the potential range of species in areas outside their natural range (Booth et al., 1989).

Quantitative approaches also have been used to investigate patterns of distribution of traits in forage legume germplasm in relation to ecogeographic features of collecting sites. Such studies have been reported for alfalfa (*Medicago sativa* L.) (Rumbaugh et al., 1988; Skinner & Stuteville, 1992; Warburton & Smith, 1993; Smith et al., 1994); birdsfoot trefoil (McGraw et al., 1989; Steiner & Poklemba, 1994); and white clover (*Trifolium repens* L.) (Daday, 1954a,b; Foulds & Grime, 1972; Caradus et al., 1990; Pederson et al., 1996). The primary analytical tools used for data exploration have been Pearson's correlation coefficient between quantitative variables, analysis of variance of the differences among classes of descriptive variables, visual examinations of multivariate clusters or principal component plots, and two-way contingency tables of qualitative groupings. These studies have provided insights into how intraspecific variation for specific traits is distributed across ranges of different environmental factors.

The four examples that follow add to this body of knowledge by integrating information from both regional and global-scale ecogeographic databases with information on plant genotypes and phenotypes based on molecular methods. These examples represent fixed inference spaces in that conclusions are limited to the sites that were sampled, and do not encompass the full ecogeographic and genetic ranges of the taxa described. However, the methods described would be applicable to broader studies of species and genotype adaptations with appropriate sampling regimes.

**Example One. Intra-Regional Ecological Distributions of Native Western Annual and Perennial *Trifolium* in the United States of America**

Multivariate factor analysis can be used as a preliminary exploration tool for defining the underlying structure of the variation in environmental variables within a collecting region. By analyzing the relationships among a set of environmental variables, it is possible to define new variables, called factors (Hair et al., 1995), that can be used to determine which of the old variables most strongly affect the ranges of species or distributions of genotypes within a species.

Germplasm collecting expeditions were conducted by Norman Taylor (Univ. Kentucky), Kenneth Quesenberry (Univ. Florida), and Warren Williams (AgResearch, New Zealand) in 1994 and 1995 to acquire accessions of native *Trifolium* species in California, Oregon, and Washington. This highly variable geographic region is rich in diversity of both annual and perennial *Trifolium* species, but only a few accessions were held and described in the USDA National Plant Germplasm System. Regions and sites where the different species were most likely to be found were identified with the assistance of local university specialists, botanical societies and herbarium specimens. Multiple accessions were collected of some species with broad distribution ranges. As few as one accession was collected of species with narrow distributions, depending upon the ability of the expedition team to locate populations. Collecting site data including latitude, longitude and elevation were recorded on collecting forms along with general descriptive notes of site habitat features.

Estimates of monthly precipitation and temperature for each collecting site were generated using meteorological station climate data for the region and a digital elevation model, the Parameter-elevation Regressions on Independent Slopes Model (PRISM; Daly et al., 1994). The collecting data for both annual and perennial species were analyzed using a common factor analysis routine with varimax rotation (SYSTAT Inc., 1992). The inference space for the ecological descriptions of species distributions is limited to the range of collection sites sampled because the expedition collecting strategy did not attempt to sample all potential sites where each species may be found. However, each site where a specimen was collected can be assumed to be representative of the kind of environment where the species may be found over a wider range of sites.

Factor analyses were carried out using climatic data for a single collecting site selected at random for each of the 21 native annual and 19 perennial *Trifolium*

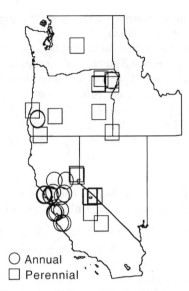

Fig. 3–2. The distribution of 23 annual and 26 perennial native *Trifolium* species collecting sites in the northwestern USA that were characterized for average monthly temperature, precipitation, elevation, latitude and longitude using common factor analysis.

species that were collected over the geographic range of the expedition (Fig. 3–2). For the annual species, the 12 average monthly temperature and precipitation values, elevation, latitude and longitude were grouped into three factors (Table 3–1A). Factor 1 included cool-season temperature (September–April), warm-season precipitation (May–September), elevation and latitude. Their grouping together means

Table 3–1. Factor analysis of climatic and geographic characteristics of: (*A*) 23 annual, and (*B*) 26 perennial native *Trifolium* species. Original rotated loading values were based on varimax rotation method. Ecogeographic variable within the same row or column are significantly associated at $P = 0.05$ and verified by Pearson's coefficient of correlation.

| Factor 1 | Factor 2 | Factor 3 |
|---|---|---|
| A. Western USA annual *Trifolium* species | | |
| Precipitation, May–September | | |
| Elevation | | |
| Latitude | | |
| Temperature, September–April | | Temperature, April & October |
| | | Temperature, May & September |
| | Longitude | |
| | Precipitation, October–April | |
| B. Western USA perennial *Trifolium* species | | |
| Precipitation, September–May | | |
| Elevation | | |
| Longitude | | |
| Temperature, November–February | Temperature, February–November | |
| | Temperature, March–October | |
| | | Precipitation, June–August |
| | | Latitude |

these variables are highly correlated. Factor 2 includes cool-season precipitation (October–April) and longitude, which are closely associated among themselves, but independent of the variable in Factor 1. Factor 3 includes average temperatures for the warm-season months (April–October).

Three factors also were identified for the perennial species (Table 3–1B). Factor 1 consisted of a narrower window of winter-month temperatures (November–February) for the perennial species than with the annuals, and also contained components describing cool-season month precipitation (September–May), elevation and longitude. Monthly average temperatures (February–November) account for the Factor 2 variation, and summer-month precipitation (June–August) and latitude for Factor 3.

Comparing the results from the annual and perennial species, it can be seen that the association of environmental variables differs for the two kinds of growth habit (Table 3–1). In particular, seasonality of precipitation and temperature distinguish the distributions of the annual and perennial species. Cool-season temperature, cool-season precipitation, and warm-season temperature are the determinant climatic features of annual species distribution. Cold-season temperature, cool-season temperature and summer precipitation are the dominant climatic features describing the range of perennial species sampled. The primary distinguishing features of ecogeographic classifications are usually precipitation and temperature, so this method provides a way to obtain more quantitative estimates of germplasm collecting site climatic effects.

## Example Two. Intra-Regional Genetic Relatedness of Western Annual and Perennial *Trifolium* Species in the United States of America

The robust clustering method described by Mantel (1967) can be used to measure the closeness of pattern distances between taxa and different kinds of environmental variables (Sokal, 1979). Distance (or similarity) measures among taxa based on genetic data (e.g., molecular marker or sequence data) can be used to construct a matrix using software packages such as PAUP (Swofford, 1993). Matrices of the actual geographic distances among collecting sites can be calculated from the great circle distances among all pairs of collecting sites. Matrices of distances based on other environmental or geographic data (e.g., monthly average temperature, precipitation, elevation) from the PRISM model results can be constructed using single or multiple combinations of collecting site characteristics. To compare the closeness of the genetic characteristics of the taxa to the environmental descriptions of the collecting sites, plots of one matrix against the other are compared element by element, with the values along the diagonal ignored, using software such as NTSYS-pc (Rohlf, 1992). The resulting Mantel Z statistic is the measure of the degree of closeness between the two matrices.

Estimates of genetic relatedness were made among *Trifolium* species in the two growth habit types by applying the Show Distance Matrix menu command in the PAUP software (Swofford, 1993) to polymorphisms in ribosomal DNA (rDNA) internal transcribed spacer (ITS) sequences (Baldwin et al., 1995). The geographic distances among collection sites within annual and perennial species can be calculated by using the latitude and longitude coordinates. Matrices of calculated Eu-

Table 3–2. Summary of Mantel Z statistic results from comparisons of matrices of collecting site ecological distances and of genetic distances from ITS sequences of 23 annual and 26 perennial native
*Trifolium* species.

| Ecovariable | Annual | Perennial |
|---|---|---|
| 12-mo temperature/precipitation pattern | ns | *** |
| Minimum temperature | ns | * |
| Maximum temperature | ns | ns |
| Total precipitation | * | ** |
| Elevation | ns | * |
| Geographic distance | ns | ns |
| Latitude | ns | *** |

clidean distances among all collection sites also were calculated using the following estimated environmental variables: average lowest month temperature, average highest month temperature, annual precipitation, elevation, geographic distance
and latitude.

These data showed that seasonal precipitation produces a pattern similar to
the genetic relationships among annual species over the range of collecting sites
used here (Table 3–2). The genetic relatedness of the perennial species, however,
is associated with a more complex array of environmental factors including coldest monthly temperature, total seasonal amounts of precipitation, as well as the geographic variables elevation and latitude. The absence of an association between
genetic similarity and geographic distance suggests that speciation may have been
due more to directional selection by environmental factors, regardless of their geographic proximity, than by the geographic proximity of genetic progenitors.

## Example Three. Distribution of Genotypes within Native *Trifolium* Species and Relationships to Environmental Variables

It is possible to seek associations of intraspecific genetic variation with ecological variables by using multiple accessions collected from different sites over
the geographic range of a species (Fig. 3–3). The Mantel Z statistic was again used
to investigate relationships between genetic and environmental similarities of collecting sites. Random amplified polymorphic DNA (RAPD) band products were
used as the indicators of genetic diversity within each species (see Steiner et al.,
1998, for general methodology).

There were no significant associations between RAPD marker diversity
within any of the annual species and the eight ecogeographic variables examined
(Table 3–3), regardless of whether the species had a limited geographic range
(e.g., *T. barbigerum*, 180-km maximum geographic distance between accessions)
or an extensive one (e.g., *T. microcephalum*, 814-km geographic distance). However, different associations with ecogeographic factors were noted for the two
perennial species. Intraspecific genetic similarity for both *T. longipes* and *T. wormskioldii* was associated with total annual precipitation. However, the genetic similarity among *T. longipes* accessions, collected across montaine regions ranging from
the southern Sierra Nevada (2200-m elevation) to the southern Oregon border
(1340-m elevation) and in an exposed rocky area near the Oregon coast (50-m elevation) (Fig. 3–3), also was associated with collecting site similarity based on el-

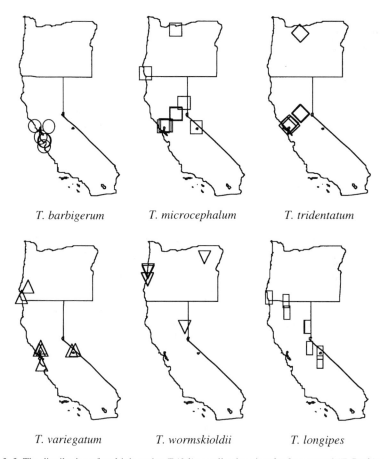

T. barbigerum          T. microcephalum          T. tridentatum

T. variegatum          T. wormskioldii          T. longipes

Fig. 3–3. The distribution of multiple native *Trifolium* collecting sites for four annual (*T. Barbigerum*, *T. microcephalum*, *T. tridentatum*, and *T. variegatum*) and two perennial (*T. Wormskioldii* and *T. longipes*) species collected in California and Oregon in 1994 and 1995.

Table 3–3. Summary of Mantel Z statistic results from comparisons of matrices of collecting site ecological distances and genetic distances based on RAPD bands from four annual and two perennial native *Trifolium* species.

| Environmental variables | Annuals† | | | | Perennials† | |
|---|---|---|---|---|---|---|
| | Barb | Micro | Trident | Var | Long | Worm |
| Temp-A | ns | ns | ns | ns | ns | ns |
| Temp-L | ns | ns | ns | ns | ns | ns |
| Temp-H | ns | ns | ns | ns | ns | ns |
| Geographic | ns | ns | ns | ns | ns | * |
| Precipitation | ns | ns | ns | ns | * | *** |
| Elevation | ns | ns | ns | ns | * | ns |
| Latitude | ns | ns | ns | ns | ns | * |
| Longitude | ns | ns | ns | ns | *** | ** |
| *n* | 6 | 9 | 7 | 11 | 8 | 9 |

† Species abbreviations: Barb = *T. barbigerum*; micro = *T. microcephalum*; Trident = *T. tridentatum*; Var = *T. variegatum*; Long = *T. longipes*; and Worm = *T. wormskioldii*.

evation and longitude. The genetic diversity of *T. wormskioldii* was associated with precipitation, latitude, and longitude. The combined effects of latitude and longitude may explain the association of *T. wormskioldii* genetic similarity with geographic distance.

These differences between annual and perennial species in how intraspecific variation is associated with ecological variables suggest that different collecting strategies may be needed to sample the range of genetic variability which also is ecologically meaningful. The absence of associations between environmental variables and annual species intraspecific genetic relatedness may indicate that a suitable number of collecting sites selected at random over the geographic range of the species will adequately represent the species. On the other hand, specific environmental differences among collecting sites can be used to identify genetic variability in perennial species.

## Example Four. Effect of Geographic Scale on Birdsfoot Trefoil Genetic and Ecogeographic Relationships

To determine the ecogeographic factors that influence the genetic diversity of birdsfoot trefoil (*Lotus corniculatus* L.) at two levels of geographic scale, accessions of the species were selected over large transcontinental (6500-km maximum geographic distance between accessions) and limited intraregional (400-km maximum distance) geographic ranges. The transcontinental range was represented by a set of 28 native birdsfoot trefoil genotypes from the National Plant Germplasm System (NPGS) birdsfoot trefoil collection selected for the geographic and ecological diversity of their collecting sites from central Russia to Atlantic Europe and from Ethiopia to Scandinavia (Steiner & Garcia de los Santos, unpublished data, 1998). The EPA/NOAA Global Ecosystems Database was used to estimate the value of environmental variables at the collecting sites. The intraregional range was represented by a set of 16 accessions collected from the western Caucasus region of Russia based on estimated precipitation and temperature differences of the region (Hart et al., 1996). Average monthly temperature and precipitation for each accession from both scale sets were used to determine Euclidean distances among the accessions. Random amplified polymorphic DNA band products were used as the indicators of genetic diversity for both plant sets using the genetic distance statistic from PAUP. The Mantel *Z* statistic was used to determine relationships be-

Table 3–4. Summary of Mantel Z statistic results from comparisons of matrices of collecting site ecological distances and of genetic distances based on RAPD bands from 28 Old World and 15 Caucasus Mountain accessions of birdsfoot trefoil.

| Ecovariable | Old World | Caucasus |
|---|---|---|
| 12-mo temperature/precipitation pattern | *** | ns |
| Minimum temperature | * | * |
| Maximum temperature | ns | ns |
| Total precipitation | ns | ns |
| Elevation | ns | * |
| Geographic distance | ns | ns |
| Latitude | ns | * |

tween intraspecific genetic variation and environmental characteristics of collecting sites.

The genetic relatedness of accessions from both the large- and small-scale geographic ranges was associated with average coldest-month temperature (Table 3–4). The genetic relatedness of accessions from the large-scale range was associated with the 12-mo pattern of precipitation and temperature, but the western Caucasus accessions were not. However, the genetic relatedness of the western Caucasus accessions was associated with elevation and latitude, while the large-scale range was not. Using this kind of information, it is possible to determine which environmental factors regulate species geographic distributions and establish at what scale one can expect to find significant intraspecific genetic variability within a limited geographic range. An approach such as this validates the germplasm collection expedition planning principles proposed by Vavilov (1992) based on a priori knowledge of patterns of environmental variation in the target region. Since there are significant differences between the two scale levels in the environmental variables that are associated with genetic relatedness of the accessions collected, care should be taken to determine which collecting site descriptors may be most appropriate to locate unique sources of genetic variation.

## CONCLUSIONS

Geographic information system tools can be used in conjunction with germplasm molecular analysis data from different species and genotypes to increase our understanding of—and improve our ability to manage—plant genetic resources. Using the kinds of methods described above, we can determine relationships between specific ecogeographic characteristics over a range of collecting sites and genetic diversity. These analyses can be used to compare the distribution patterns of genetic variation at different levels of geographic scale and to help plan future collection expeditions so that novel germplasm is obtained. By integrating what we already know about germplasm holdings with information obtained from molecular and GIS technologies, the gene bank manager can help ensure that these valuable resources are efficiently utilized, and thus ensure that valuable traits are preserved.

## REFERENCES

Bailey, R.G. 1984. Testing an ecosystem regionalization. J. Environ. Manage. 19:239–248.

Bailey, R.G. 1989. Explanatory supplement to ecoregions map of the continents. Environ. Conserv. 16:307–309.

Baldwin, B.G., M.J. Sanderson, J.M. Porter, M.F. Wojciechowski, C.S. Campbell, and M.J. Donoghue. 1995. The ITS region of nuclear ribosomal DNA: A valuable source of evidence on angiosperm phylogeny. Ann. Missouri Bot. Garden 82: 247–277.

Booth, T.H., J.A. Stein, H.A. Nix, and M.F. Hutchinson. 1989. Mapping regions climatically suitable for particular species: An example using Africa. For. Ecol. Manage. 28:19–31.

Caradus, J.R., A.C. MacKay, J.F.L. Charlton, and D.F. Chapman. 1990. Genecology of white clover (*Trifolium repens* L.) from wet and dry hill country pastures. N.Z. J. Agric. Res. 33:377–384.

Chrtkova-Zertova, A. 1973. A monographic study of *Lotus corniculatus* L. in central and northern Europe. Rozpravy Ceskoslov. Akead. Ved Praha 83:1–94.

Daday, H. 1954a. Gene frequencies in wild populations of *Trifolium repens*. I. Distribution by latitude. Heredity 8:61–78.

Daday, H. 1954. Gene frequencies in wild populations of *Trifolium repens*. II. Distribution by altitude. Heredity 8:377–384.

Daly, C., R.P. Neilson, and D.L. Phillips. 1994. A statistical-topographic model for mapping climatological precipitation over mountainous terrain. J. Appl. Meteorol. 33:140–158.

Duke, J.A. 1978. The quest for tolerant germ plasm. p. 1–61. *In* G.A. Jung (ed.) Crop tolerance to suboptimal land conditions. ASA, CSSA, SSA, Madison, WI.

Foulds, W., and J.P. Grime. 1972. The response of cyanogenic and acyanogenic phenotypes of *Trifolium repens* to soil moisture supply. Heredity 28:181–187.

Greene, S.L., and J.R. McFerson. 1994. Conservation of Lotus genetic resources: Status of the U.S. collection. p. 39–42. *In* P.R. Beuselinck and C.A. Roberts (ed.) Proc. 1st Int. Lotus Symp., St. Louis, MO. 22–24 March. Univ. Missouri Exten., Columbia, MO.

Hart, T.C., S.L. Greene, and A. Afonin. 1996. Use of ecological and geographical databases for prognostic intraregional plant distribution mapping. p. 6–8. *In* D. Geltman and Y. Roskov (ed.). Computer databases in botanical research. (In Russian.) Konorov Bot. Inst., St. Petersburg, Russia.

Mantel, N. 1967. The detection of disease clustering and a generalized regression approach. Cancer Res. 27:209–220.

Pederson, G.A., T.E. Fairbrother, and S.L. Greene. 1996. Cyanogenesis and climatic relationships in U.S. white clover germ plasm collection and core subset. Crop Sci. 36:427–433.

Pistorius, R. 1997. Scientists, plants and politics—a history of the plant genetic resources movement. Int. Plant Genet. Resour. Inst., Rome, Italy.

Radloff, D.L., and D.R. Betters. 1978. Multivariate analysis of physical site data for wildland classification. Forest Sci. 24:2–10.

Rick, C.M. 1973. Potential genetic resources in tomato species: Clues from observations in native habitats. p. 255–270. *In* A.M. Srb (ed.) Genes, enzymes and population. Vol. 2. Plenum Press, New York.

Rohlf, F.J. 1992. NTSYS-pc numerical taxonomy and multivariate analysis system. Version 1.7 (computer software and manual). Exeter Software, Setauket, NY.

Rumbaugh, M.D., W.L. Graves, J.L. Caddel, and R.M. Mohammad. 1988. Variability in a collection of alfalfa germ plasm from Morocco. Crop Sci. 28:605–609.

Skinner, D.Z., and D.L. Stuteville. 1992. Geographical variation in alfalfa accessions for resistance to two isolates of *Peronospora trifoliorum*. Crop Sci. 32:1467–1470.

Smith, S.E., D.W. Johnson, D.M. Conta, and J.R. Hotchkiss. 1994. Using climatological, geographical, and taxonomic information to identify sources of mature-plant salt tolerance in alfalfa. Crop Sci. 34:690–694.

Sokal, R.R. 1979. Testing statistical significance of geographic variation patterns. Syst. Zool. 28:227–232.

Steiner, J.J. 1994. Lotus germ plasm utilization: integrating genetic diversity, species relationships, and ecological distributions. p. 43–50. *In* P.R. Beuselinck and C.A. Roberts (ed.) Proc. 1st Int. Lotus Symp., St. Louis, MO. 22–24 March. Univ. Missouri Ext., Columbia, MO.

Steiner, J.J. 1999. Birdsfoot trefoil origins and germ plasm diversity. p. 39–50. *In* P.R. Beuselinck (ed.) Trefoil: The science and technology of *Lotus*. CSSA Spec. Publ. 27. Madison, WI. (In press.)

Steiner, J.J., and S.L. Greene. 1996. Proposed ecological descriptors and their utility for plant germ plasm collections. Crop Sci. 36:439–451.

Steiner, J.J., E. Piccioni, M. Falcinelli, and A. Liston. 1998. Germ plasm diversity among cultivars and the NPGS crimson clover collection. Crop Sci. 38:263–271.

Steiner, J.J., and C.J. Poklemba. 1994. *Lotus corniculatus* classification by seed globulin polypeptides and relationship to accession pedigrees and geographic origin. Crop Sci. 34:255–264.

Swofford, D.L. 1993. PAUP: Phylogenetic analysis using parsimony, Ver. 3.1.1 computer program. Illinois Natural Hist. Surv., Champaign, IL.

SYSTAT Inc. 1992. SYSTAT: Statistics. Version 5.2 ed. SYSTAT, Inc., Evanston, IL.

Upvardy, M.D.F. 1975. A classification of the biogeographical provinces of the world. IUCN Occasional Pap. 18. IUCN, Morges, Switzerland.

Vavilov, N.I. 1992. Problems concerning new crops. p. 256–285 *In* V.F. Dorofeyev (ed.) N.I. Vavilov: Origin and geography of cultivated plants. Univ. Press, Cambridge, UK.

von Bothmer, R., and O. Seberg. 1995. Strategies for the collecting of wild species. p. 93–111. *In* L. Guarino et al. (ed.) Collecting plant genetic diversity. CAB Int., Wallingford, UK.

Warburton, M.L., and S.E. Smith. 1993. Regional diversity in nondormant alfalfas from India and the Middle East. Crop Sci. 33:852–858.

Zobler, L. 1957. Statistical testing of regional boundaries. Ann. Assoc. Am. Geogr. 47:83–95.

# 4

# Scale Considerations in Mapping for Germplasm Acquisition and the Assessment of Ex Situ Collections

**Thomas C. Hart**

*Spatial Data Associates*
*Trumensburg, New York*

Spatial data on habitats can guide as well as inform the selection of sites for plant germplasm acquisition and/or in situ conservation and the evaluation of accessions conserved ex situ (Guarino et al., 1999, see Chapter 1). They help in the search for target habitats, and can give a good overview of the conditions under which the populations represented in a collection were growing (Jones et al., 1997). Most plant genetic resource (PGR) workers use spatial data of some sort in their work, and most would forecast a substantial increase in the use of digital spatial data sets in the coming decade, as digitization reaches farther and more strongly across both geographic areas and topics of interest to germplasm conservation programs.

However, an intractable problem must be overcome. Most habitat factors that are important to PGR conservationists and users are far more intricate than can be mapped consistently across broad areas. The conditions that link habitat character to adaptive traits in germplasm are often so limited in spatial area that mismatches are found when comparisons are drawn using the extensive descriptors of map products. The following discussion will explore the implications of this mismatch, and will offer some conclusions dealing more with its acceptance than its solution.

Closely associated with this mismatch are the ongoing changes in the nature of map quality and in the meaning of scale, particularly with respect to the size of depiction and the inconsistency with which map data are delivered in an age of do-it-yourself digital mapping. A thorough review of the challenges of map scale in a digital world is provided by Goodchild and Proctor (1997). Additional useful orientation to the field scientist's perspective on issues of scale can be found in Carlile et al. (1989) and Turner et al. (1989).

This chapter will examine considerations of scale by way of the opportunities and constraints presented by a specific field exploration and a series of subsequent analyses of both new and long-established collections. The examples are drawn from products prepared for the 1995 West Caucasus Collecting Expedition (Fig. 4–1), a collaboration between the U.S. Department of Agriculture (USDA)

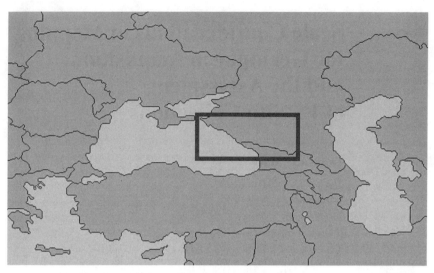

Fig. 4–1. Study region for 1995 USDA/VIR collecting expedition for legume germplasm in the West Caucasus region of southern Russia (from Greene et al., 1999a).

and Russia's Vavilov Institute for Plant Industry (VIR) in St. Petersburg, which sought germplasm of forage legume taxa adapted to acidic soil conditions. A Geographic Information System (GIS) study was sponsored by the USDA's National Plant Germplasm System (NPGS) in support of the joint collecting program. The map products were compiled using software toolkits from ERDAS Imagine NT (ERDAS Inc., Atlanta, GA), ArcInfo (ESRI Inc., Redlands, CA), Surfer (Golden Software Inc., Golden, CO), Excel (Microsoft Inc., Bellevue, WA) and IDRISI (Graduate School of Geography, Clark Univ., Worcester, MA) the latter two serving as distribution formats due to their broad user base. A detailed description of the methodology used in this project can be found in Greene et al. (1999a)

## THE CHANGING DEFINITION OF SCALE

In precomputerized traditional mapping environments, scale was a simple concept that referred to the ratio between the extent of the map and the ground area to which it corresponded. Typically, this ratio was shared by a standard map series, so that users could refer to scale as a summary for map detail and reliability. Scale became a primary guide to appropriate data usage, with accepted thematic content and expectations of positional reliability for each level on the gradient between local and broad (Raisz, 1962; Aberley, 1993).

With digital cartography, many of the reassurances of scale have disappeared. First, maps stored in electronic form can theoretically be displayed at any size, so that the original sense of scale needs to be replaced with a new concept, perhaps "spatial resolution" or "minimum mapping unit" (Chrisman, 1997, p. 77–83; Malanson & Armstrong, 1997). This quandary is illustrated in simple form

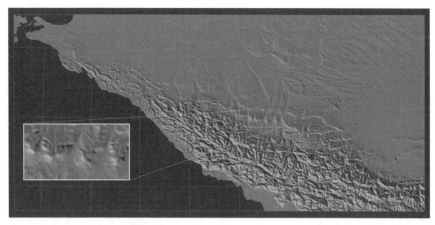

Fig. 4–2. West Caucasus shaded relief map with inset to illustrate the duplication of details in digital data regardless of the degree of enlargement.

by Fig. 4–2, where the inset, although enlarged, contains the same degree of shaded relief detail as does the main map. A second source of scale confusion in digital maps is their hybrid character. They are often compiled from a variety of sources that are of mixed original scale, or of an undefined scale. Examples include an interpolation of point records or boundary data derived from scanner imagery (Berry, 1994). Finally, the most prominent reason for the decay of scale as an indicator of a map's quantity and quality of detail lies in the accelerating increase in the number and diversity of map makers. Map-making standards have declined, as has the consistency of methodology and symbology (Monmonier, 1995). This is probably short-lived, as the scientific world will soon implement fresh standards for digital mapping, as well as an effective set of indications of adherence (Univ. Consort. Geogr. Inform. Sci., 1996).

## SCALE SUFFICIENCY FOR GENETIC RESOURCES RESEARCH

Whether or not another term replaces "scale" as the generic descriptor of map quality, there remain some key measures of the usefulness of a map's data for germplasm projects. These can be summarized as predictable levels of accuracy and of precision for the map parameters of position and legend category. In reference to their accuracy, both position and category can be wrong or right, and in reference to their precision either can be depicted in a generalized or a highly specific manner. As a rule, there is a trend for more precision in map data from more local scales, for both position and category. A weaker but similar correlation applies to accuracy.

Perhaps the most elusive aspect of map quality relates to the consistency with which small units are handled. Pockets of distinct character either between or within larger units often get prominent treatment when they are easily discerned, or when they represent radical categories of particular interest to those who compiled the original data. However, they also are often selectively overlooked, because there is already sufficient local activity in the map, or their categorical makeup allows lumping with neighboring areas. These small anomalous units can be very important from a germplasm conservation perspective, either due to their unique habitat or because they harbor cut-off "island" populations in vestige habitats. Their importance will depend on the target taxon and on the biogeographic history of the area, such as recent patterns of disturbance and the degree of habitat fragmentation.

It can be expected that where scales are more local, the map treatment of these small areas becomes more predictable, and more faithful to their unique characteristics. However, map data rarely characterize in sufficient detail a major portion of the small isolated patches where populations of the target taxon offer appropriate germplasm collecting sites. There are too many special circumstances such as ravines, cliff edges or localized soil and/or slope/aspect conditions, to warrant much confidence in spatial data as the primary site descriptors. In the habitat shown in Fig. 4–3 special mesic conditions were observed at the base of the cliff. But a move in position of as little as 10 m results in significant change in moisture and temperature regimes. This type of small isolated patch is probably impossible to capture as a consistent feature on an extensive map, but this does not diminish its importance as potential target habitat for germplasm sampling.

There is therefore a lasting need for diligent onsite observation and for the documentation of site ecological data using standardized formats at the time of collection. These are typically attributes that the germplasm scientist is accustomed to gathering and interpreting. By comparing the field data to the spatial data, the collector perceives connections across the gap in scale, so that there is a better application of the map at the local level, and conversely a better context in which to

Fig. 4–3. Locally varied habitat on margins of a small cliff in the center of the West Caucasus Russia study region.

evaluate direct local observations. As an outcome of the process, the collector is able to determine how well each site is represented by its map neighborhood, and can decide whether to assign map data (such as climatic attributes) directly to the site in the case of good fit, or to withhold the map descriptors in the case of an un-mapped inlier or other anomalous circumstance.

This process is illustrated in Fig. 4–4. The 1995 West Caucasus samples of *Lotus corniculatus* L. (birdsfoot trefoil) are listed in three groups (sorted vertically as white, light gray, and medium gray) according to the quantity of seed collected, and are described in the "Selection" column if there are anomalous site conditions relative to the mapped neighborhood. Such "flagging" would be very useful during germplasm evaluation in identifying populations that may have highly specific adaptations compared to their neighbors, or that may have been isolated from other populations and likely to have diverged genetically.

**49 Seed Samples: Experimental criteria for correlating habitat with genetic measures and adaptation**

Lotus comiculatus

| Date | Utm E | Utm N | Acc# | S# | pH | Elev | Slope | Aspect | Grad Pos | Selection | Tz | N11.2 | Wtemp | Mz | Panngrp | Pann | PJan | PFeb | PMar | 55 more climate attributes --> |
|---|---|---|---|---|---|---|---|---|---|---|---|---|---|---|---|---|---|---|---|---|
| 950907 | 716892 | 4817055 | 269 | 99 | 6.0 | 1864 | 13 | 5 | t3w m5 | hwy graze | 12 | -27 | 3 | 7 | 5 | 1716 | 167 | 164 | 144 | |
| 950907 | 716892 | 4817055 | 270 | 98 | 6.0 | 1864 | 13 | 5 | t3w m5 | | 12 | -27 | 3 | 7 | 5 | 1716 | 167 | 164 | 144 | |
| 951003 | 566754 | 4877850 | 342 | 127A | 7.0 | 1366 | 19 | 7 | t3c m5 | 5med | 12 | -28 | 3 | 6 | 5 | 1728 | 186 | 169 | 159 | |
| 950907 | 718268 | 4813319 | 261B | 102 | 5.5 | 2296 | 19 | 3 | t3(a) m5 | 5medac | 13 | -28 | 3 | 6 | 5 | 1798 | 176 | 175 | 154 | |
| 960909 | 729603 | 4793014 | 357B | 106 | 5.9 | 2113 | 27 | 6 | t2 m5 | 5warm | 12 | -25 | 2 | 6 | 5 | 1833 | 181 | 175 | 155 | |
| 950921 | 574894 | 4891256 | 308 | 115 | 6.5 | 1291 | 6 | 8 | t4 m4 | 4cold | 10 | -31 | 4 | 8 | 4 | 1324 | 122 | 119 | 114 | |
| 950922 | 582744 | 4878020 | 315 | 117 | 7.5 | 1868 | 17 | 2 | t4c m4 | 4vcold | 13 | -33 | 4 | 7 | 4 | 1494 | 140 | 138 | 133 | |
| 950829 | 585727 | 4891955 | 169 | 82 | 7.0 | 1310 | 7 | 8 | t4c m3 | ripar, rds | 10 | -33 | 4 | 9 | 3 | 1111 | 84 | 89 | 90 | |
| 950901 | 582418 | 4890002 | 199 | 88 | 7.0 | 1281 | 1 | 0 | t4c m3 | 3vcold | 10 | -32 | 4 | 8 | 3 | 1182 | 98 | 99 | 98 | |
| 950901 | 582418 | 4890002 | 207 | 88 | 7.0 | 1291 | 1 | 0 | t4 m3 | | 10 | -32 | 4 | 8 | 3 | 1182 | 98 | 99 | 98 | |
| 950914 | 595695 | 4889371 | 285 | 108 | 5.5 | 1085 | 14 | 3 | t4(a) m3 | 3coldac | 10 | -33 | 4 | 10 | 3 | 956 | 58 | 64 | 69 | |
| 950919 | 575329 | 4933099 | 301 | 112 | 5.2 | 368 | 3 | 6 | t4(a) m3 | 3coldac | 8 | -32 | 4 | 3 | 3 | 905 | 70 | 69 | 73 | |
| 950919 | 575329 | 4933099 | 302 | 112 | 5.2 | 368 | 3 | 6 | t4(a) m3 | | 8 | -32 | 4 | 3 | 3 | 905 | 70 | 69 | 73 | |
| 950921 | 585753 | 4887512 | 313 | 116 | 6.8 | 1141 | 8 | 4 | t4w m3 | 3cold | 10 | -31 | 4 | 9 | 3 | 1116 | 88 | 88 | 88 | |
| 950922 | 587395 | 4895221 | 321 | 119 | 7.2 | 1197 | 6 | 8 | t4c m3 | 3vcold | 10 | -33 | 4 | 9 | 3 | 1039 | 73 | 81 | 82 | |
| 950816 | 498059 | 4949246 | 78 | 61 | 5.2 | 190 | 1 | 0 | t3(a) m3 | 3medac | 6 | -29 | 3 | 3 | 3 | 929 | 89 | 87 | 82 | |
| 950924 | 572515 | 4911104 | 326 | 121 | 6.6 | 451 | 2 | 0 | t3 m3 | ravine | 6 | -30 | 3 | 9 | 3 | 1076 | 93 | 88 | 87 | |
| 950810 | 573793 | 4915117 | D65 | 50 | 7.7 | 390 | 5 | 4 | t3w m3 | disturbed | 6 | -30 | 3 | 9 | 3 | 1021 | 85 | 81 | 82 | |
| 950723 | 684080 | 4906136 | 37 | 29 | 7.5 | 783 | 1 | 0 | t4 m2 | roadway | 8 | -33 | 4 | 11 | 2 | 773 | 42 | 41 | 47 | |
| 950724 | 693572 | 4921341 | 41 | 31 | 7.8 | 563 | 4 | 4 | t4 m2 | 2cold | 8 | -32 | 4 | 13 | 2 | 653 | 34 | 30 | 38 | |
| 950803 | 610038 | 4914083 | 66 | 47 | 3.9 | 525 | 4 | 4 | t4(a+) m2 | 2coldac+ | 8 | -31 | 4 | 13 | 2 | 720 | 26 | 36 | 48 | |
| 950805 | 598002 | 4920647 | D78 | 48 | 7.8 | 388 | 14 | 6 | t4w m2 | mowed | 6 | -31 | 4 | 11 | 2 | 761 | 38 | 45 | 54 | |
| 950725 | 727669 | 4899979 | 43 | 33 | 7.4 | 688 | 5 | 3 | t3 m2 | landslide | 8 | -30 | 3 | 11 | 2 | 770 | 43 | 41 | 44 | |
| 950820 | 416969 | 4948291 | 112 | 70 | 6.9 | 469 | 20 | 7 | t2c m2 | 2warm | 3 | -23 | 2 | 2 | 2 | 725 | 76 | 75 | 62 | |
| 950729 | 824468 | 4819835 | 57 | 42 | 8.5 | 1063 | 9 | 4 | t3w m1 | orch terrs | 9 | -27 | 3 | 14 | 1 | 585 | 15 | 17 | 30 | |
| 950730 | 668952 | 4976302 | 61 | 44 | 7.4 | 219 | 3 | 3 | t3 m1 | quarry | 5 | -30 | 3 | 13 | 1 | 628 | 42 | 38 | 43 | |
| 950709 | 339579 | 5025261 | 3 | 3 | 7.2 | 52 | 1 | 0 | t2 m1 | quarry | 3 | -23 | 2 | 1 | 1 | 473 | 45 | 46 | 35 | |
| 950711 | 379256 | 4989736 | 16 | 10 | 7.2 | 59 | 6 | 4 | t2 m1 | 1warm | 3 | -24 | 2 | 1 | 1 | 591 | 59 | 55 | 48 | |
| 951003 | 566754 | 4877850 | 341 | 127A | 7.0 | 1366 | 19 | 7 | t3c m5 | | 12 | -28 | 3 | 6 | 5 | 1728 | 185 | 169 | 159 | |
| 950922 | 581603 | 4884425 | 320 | 118 | 6.4 | 1471 | 6 | 7 | t4c m4 | 4cold | 10 | -32 | 4 | 8 | 4 | 1311 | 116 | 115 | 112 | |
| 950905 | 647634 | 4882997 | 232 | 94 | 5.4 | 659 | 8 | 7 | t4(a) m3 | 3coldac | 8 | -32 | 4 | 11 | 3 | 860 | 32 | 38 | 49 | |
| 950914 | 595695 | 4889371 | 279 | 108 | 5.5 | 1085 | 14 | 3 | t4(a) m3 | | 10 | -33 | 4 | 10 | 3 | 956 | 58 | 64 | 69 | |
| 950919 | 574438 | 4905545 | 295 | 111 | 5.2 | 408 | 3 | 6 | t4(a) m3 | | 8 | -31 | 4 | 3 | 3 | 969 | 76 | 75 | 78 | |
| 950920 | 586101 | 4896204 | 306 | 114 | 6.2 | 1163 | 11 | 2 | t4 m3 | 3cold | 10 | -33 | 4 | 9 | 3 | 1042 | 74 | 82 | 84 | |
| 950922 | 591172 | 4896272 | 323 | 120 | 7.6 | 772 | 13 | 5 | t4w m3 | 3cold | 9 | -31 | 4 | 10 | 3 | 902 | 54 | 62 | 68 | |
| 950919 | 569003 | 4911499 | 292 | 110 | 6.0 | 408 | 7 | 6 | t3 m3 | 3med | 6 | -30 | 3 | 4 | 3 | 1119 | 101 | 94 | 93 | |
| 950923 | 555926 | 4945048 | 155 | 72 | 7.7 | 363 | 9 | 6 | t1 | hi rd traffic | 3 | -13 | 2 | | 5 | 2038 | 250 | 237 | 190 | |
| 950831 | 581731 | 4878961 | 189 | 85 | 4.6 | 1871 | 4 | 8 | t4(a+)m4 | 4coldac+ | 13 | -32 | 4 | 8 | 4 | 1493 | 140 | 139 | 133 | |
| 950906 | 709110 | 4844035 | 245 | 96 | 5.2 | 1207 | 3 | 2 | t3(a) m4 | 4medac | 9 | -27 | 4 | 8 | 4 | 1415 | 136 | 124 | 112 | |
| 950812 | 519364 | 4933679 | D109 | 60 | 6.2 | 188 | 6 | 3 | t3 m4 | passport? | 6 | -30 | 3 | 4 | 4 | 1301 | 130 | 116 | 106 | |
| 950811 | 560136 | 4900658 | D93 | 53 | 5.4 | 338 | 3 | 3 | t3(a) m4 | passport? | 6 | -30 | 3 | 4 | 4 | 1342 | 139 | 121 | 114 | |
| 950921 | 585753 | 4887512 | D170 | 116 | 6.8 | 1141 | 8 | 4 | t4w m3 | | 10 | -31 | 4 | 9 | 3 | 1116 | 88 | 88 | 88 | |
| 950922 | 587395 | 4895221 | D172 | 119 | 7.2 | 1197 | 6 | 8 | t4c m3 | | 10 | -33 | 4 | 9 | 3 | 1039 | 73 | 81 | 82 | |
| 950919 | 569003 | 4911499 | 294 | 110 | 6.0 | 408 | 7 | 6 | t3 m3 | | 6 | -30 | 3 | 4 | 3 | 1119 | 101 | 94 | 93 | |
| 950821 | 449441 | 4934160 | 126 | 71 | 5.6 | 675 | 9 | 3 | t2 m3 | 3warm | 3 | -24 | 2 | 2 | 3 | 923 | 99 | 98 | 81 | |
| 950918 | 604388 | 4903802 | 286 | 109 | 6.1 | 799 | 9 | 4 | t4w m2 | 2cold | 8 | -32 | 4 | 11 | 2 | 784 | 32 | 44 | 54 | |
| 950817 | 418109 | 4973664 | D123 | 65 | 7.7 | 39 | 6 | 4 | t3 m2 | 2med | 5 | -30 | 3 | 3 | 2 | 726 | 70 | 74 | 62 | |
| 950729 | 768798 | 4835603 | D49 | 34 | 7.7 | 740 | 7 | 5 | t3 m2 | 2med | 9 | -28 | 3 | 13 | 2 | 719 | 34 | 34 | 37 | |
| 950731 | 629216 | 4945628 | D73 | 46 | 7.5 | 300 | 2 | 0 | t3 m2 | riparian | 4 | -31 | 3 | 13 | 2 | 729 | 36 | 46 | 49 | |

aspect 1 = N .... 8 = NW

Seed volume per sample (row):
  three experiments
  two experiments
  one experiment

Selection subsets:
  (row) less useful of a pair of site samples
  best for moisture / temperature tests
  too close to other sites
  flawed by local unmapped factors

Grad Pos codes: t(warm to cold), m(dry to wet)
  suffix warm or cold due to aspect/slope
  pH shown below 5.5 as "a", below 5.0 as "a+"

Selection codes: moisture 1-5 (dry to wet)
  with temp descriptor and acidity if pH < 5.5

Fig. 4–4. Site and seed lot characteristics for the expedition's collection of *Lotus corniculatus* accessions, prioritized by the amount of seed collected, habitat, spatial autocorrelation and representation of map information.

## THE ADVANTAGES OF INVESTING IN SCALE COMBINATIONS

Spatial data that approximate population-level description are seldom available for key habitat descriptors, but nevertheless most conservation projects will tend to give higher importance to more detailed data. When this demand for maximum detail is combined with the size of the typical area of investigation, a serious problem of excessive data volume results, in terms of ease of storage, exchange, access and manipulation. Even when data quality falls well short of ideal, there typically will be a very large body of mapping data to process.

The costs involved in compiling a detailed, comprehensive and reasonably consistent set of habitat indicators for a study area often appear out of proportion to the direct research contribution, especially when judged from the perspective of longstanding practice. Detailed spatial data sets can be expensive to obtain, and the time commitment required to compile the data are often large relative to the other tasks involved in the preparation of germplasm conservation projects. In addition, a query of map data often eliminates vast portions of the compiled material, leading to an impression of wastage. Also, after extensive adjustment, the mapped results may well turn out to be inconclusive or irrelevant, either because the variation is too complex, or not complex enough, or because of a lack of linkage between habitat and plant genetic differences for the taxa under study.

Compensating for these disappointments are many benefits, both obvious and subtle. The most elementary advantage of compiling spatial data is that they can be manipulated to allow the conservationist to step away from the local detail of a study area and view its broad patterning. Wider trends can be assessed and differences in variability between subregions summarized (Quattrochi & Goodchild 1997), providing key inputs into the design and adjustment of sampling plans for both ex situ and in situ conservation. An array of maps in a scale series will show patterns of agreement and divergence: the salient and robust characteristics will be corroborated through the set, while obscure or inconsistent ones will be refuted by failing to appear in each map. The particulars of nesting patterns as one compares a sequence of scales are excellent introductions to a bioregional understanding (e.g., the geometry of orographic climatic patterns), which can then serve as a framework for subsequent examinations. There is rarely disappointment about the time spent stepping away to a broad scale for orientation before refocusing at the local level.

Two time-proven techniques allow the compositing of different scales and provide unique leverage for the analysis of spatial data sets: interpolation and stratification. These are examples of how spatial data can be manipulated to provide levels of understanding unreachable by the mere examination of maps. Each in its particular way mitigates against the limitations of original map scale.

Interpolation involves the calculation of a grid of new values from the data provided at a limited number of map positions, expanding the detail from points into the "gaps" between the points, in effect rendering the data at a more local scale. This is most commonly undertaken for climatic attributes collected at long-term meteorological stations, but is equally effective for other point-sampled data whose variation over space and time tends to be relatively gradual. A large boost in effectiveness is achieved by using covariant data (such as elevation in the case of climate) to strengthen the interpolation, so that intervening local features can exhibit more realistic behaviors. In Fig. 4–5, the upper right quadrant shows long-term May precipitation, a variable found to be correlated positively with elevation as mapped in 500-m cells. The resulting climate interpolation shows strong terrain patterns, which were however lessened when data from time periods whose precipitation is less well correlated with elevation were used. Nonetheless, for the case of 113 meteorological stations in the West Caucasus, it was certainly worthwhile to undertake the extra task of incorporating elevation into the interpolation of climate data (Hart et al., 1996; Hutchinson, 1995).

Stratification is the lumping of map detail into larger contiguous areas, so that from divergent inputs may emerge a simplified overview with fewer classes and a more easily interpretable pattern. This is achieved by logical or mathematical reductions of map layers, and can be considered a transformation into broader scale information. In Fig. 4–5, the lower right quadrant shows the result of clustering on 15 seasonal attributes of moisture and temperature to create subregional strata. This had two purposes: to encourage a reasonably complete sampling across the gradients offered by the study area, and to understand the distribution of climate in a transitional area between major continental and maritime climatic regimes (Hart et al., 1996; Greene et al., 1999a). The effect of multiple blendings is seen in the contrast between the upper right's subtleties and the lower right's lobate patterns.

Some limitations of the combination approach also are seen in Fig. 4–5. The climate attributes were compiled without regard to terrain aspect, a clear shortcoming when attempting to characterize plant microhabitat. However, the patterns of aspect shown in the lower left are so complex that to combine aspect with the clustered climate zones would have produced maps so convoluted the field team would not have had sufficiently simple indications on which to make decisions. Furthermore, the vast majority of the collecting sites were to be found in meadows that have faded pink signatures on the false color imagery in the upper left, and these were limited enough in frequency and extent that a wholesale recombination of all maps into a master strata system was unnecessary. In the field, it was easy first to locate meadows and then to evaluate whether their climate and aspect attributes made them strong candidates for sampling. In summary, combinations of data were useful as long as they avoided excessive fragmentation of the study area, and also as long as there resulted a categorically stable and sensible aggregation. The alternative to combining data was a simultaneous examination of the various maps at their original detail and graphic optimization, which often proved to be the most informative approach.

## CONCLUSIONS

The decision as to whether to attempt to use digital methods to analyze spatial data in PGR exploration work involves many judgements across a spectrum of issues: including project resources, taxa of interest, complexity of the study area, and quality of existing maps (Greene & Hart, 1999, see Chapter 2). After considering the costs, and deciding there is some linkage between important traits of target taxa and the habitats within the study area, the key indicator that digital spatial compilation will be worthwhile is that the texture of the landscape matches rea-

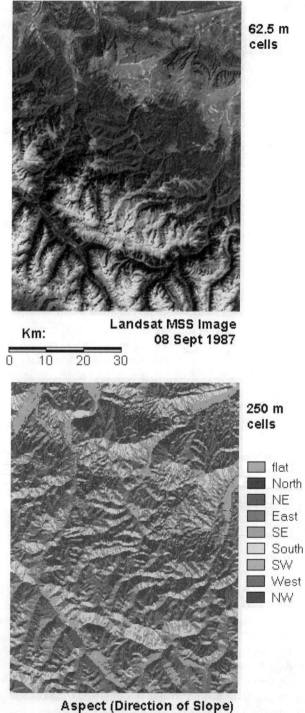

Fig. 4–5. Four GIS layers of the same area with different cell sizes and patterns, illustrating the interpretive value of nested multi-scale map data (from Greene et al., 1999a).

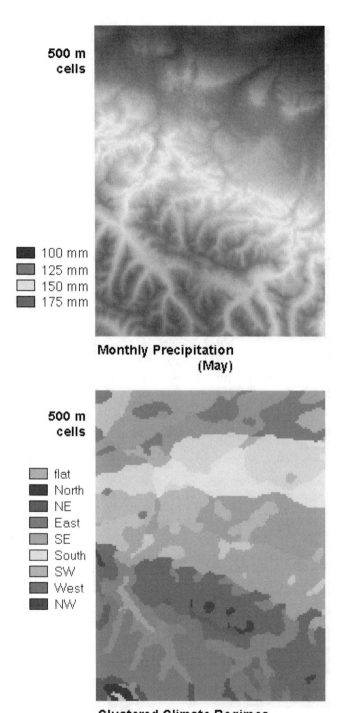

500 m
cells

■ 100 mm
■ 125 mm
□ 150 mm
■ 175 mm

Monthly Precipitation
(May)

500 m
cells

□ flat
■ North
■ NE
■ East
□ SE
□ South
□ SW
■ West
■ NW

Clustered Climate Regimes
(Moisture + Temperature)

Fig. 4–5. Continued

sonably well with the expected quality of the supporting maps. If the landscape is too complex relative to spatial data of affordable quality, then there will be discontinuity between the pattern of map data and the typical patch size. If, on the other hand, the study area is so simple as to be mapped easily using traditional tools, then the effort and expense may not pay off. In between, there are plenty of opportunities for digital spatial data to make important contributions.

However, even in the best of applications, there will remain an annoying frequency of mismatches between map and ground. There always will be attributes that can only be measured with certainty on site, and maps will never replace the central data element of consistent and reducible site observation. Conversely, there are usually data needs that cannot be handled well without maps, such as distance measures, access questions, and attributes such as long-term climate that are integrated with respect to space and/or time.

Finally, a useful caution is that one should never expect complete satisfaction from spatial data. At best they are approximations, are frequently modified by special circumstances that inhibit consistency, and are usually lacking verification. If their compilation and analysis are judged to be too much trouble relative to the potential benefit, it also is usually the case that for those closest to the scientific decisions on a project, detailed digital analysis of map data results in unexpected benefits (for example, Greene et al., 1999b). This is due in part to the subtle learning that has been encouraged by the spatial patterning, but even more important to the sense of how the data failed. There is great value in gaining familiarity with what is *not* known, as well as with what is known. Spatial analysis can provide a useful framework for an assessment of what remains to be understood about the relationship between a landscape and the genetic resources that may be found there.

## REFERENCES

Aberley, D. 1993. Boundaries of home. The New Catalyst Bioregion. Ser. New Soc. Publ., Philadelphia, PA.

Berry, J.K. 1994. Empirical verification assesses mapping performance. GIS World 7(12):28.

Carlile, D.W., J.R. Skalski, J.E. Batker, J.M. Thomas, and V.I. Cullinan, 1989. The determination of ecological scale. Landscape Ecol. 2:203–213.

Chrisman, N. 1997. Exploring geographic information systems. John Wiley & Sons, Inc., New York.

Goodchild, M.F., and J. Proctor. 1997. Scale in a digital geographic world. Geograph. Environ. Model. 1:5–23.

Greene, S.L., T.C. Hart, and A. Afonin 1999a. Using geographic information to acquire wild crop germplasm for *ex situ* collections: I. Map development and use. Crop Sci. 39:836–842.

Greene, S.L., T.C. Hart, and A.Afonin. 1999b. Using geographic information to acquire wild crop germplasm for *ex situ* collections: II. Post collection analysis. Crop Sci. 39:843–849.

Hart, T.C., S.L. Greene, and A. Afonin, 1996. Mapping for germplasm collections: Site    selection and attribution. *In* Proc. 3rd Int. Conf. Integrat. GIS and Environ. Model., Santa Fe, NM. 21–25 January. NCGIA, Santa Barbara, CA.

Hutchinson, M.F. 1995. Interpolating mean rainfall using thin plate smoothing splines. Int. J. GIS 9:305–403.

Jones, P.G., S.E. Beebe, and J. Tohme. 1997. The use of geographical information systems in biodiversity exploration and conservation. Biodivers. Conserv. 6:947–958.

Malanson, G.P., and M.P. Armstrong, 1997. Issues in spatial representation: Effects of number of cells and between-cell step size on models of environmental processes. Geograph. Environ. Model. 1:47–64.

Monmonier, M.. 1995. Drawing the line: tales of maps and cartocontroversy. Henry Holt and Co., New York.

Raisz, E. 1962. Principles of cartography. McGraw-Hill, New York.

Quattrochi, D.A., and M.F. Goodchild, 1997. Scale in remote sensing and GIS. Lewis Publ. / CRC Press, Boca Raton, FL.

Turner, M.G., R.V. O'Neill, R.H. Gardner, and B.T. Milne. 1989. Effects of changing spatial scale on the analysis of landscape pattern. Landscape Ecol. 3(3/4):153–162.

University Consortium for Geographic Information Science. 1996. Research priorities for geographic information science. Cartogr. Geogr. Inform. Syst. 23(3):115–127.

# 5    Matching Germplasm to Geography: Environmental Analysis for Plant Introduction

Trevor H. Booth

*CSIRO Forestry and Forest Products*
*Kingston, Canberra, Australia*

To satisfy the needs of the world's rapidly expanding population, the Earth's finite resources must be used with increasing efficiency. Many countries need to reforest areas to reduce land degradation and to meet increasing demand for wood and paper products. The choice of tree species and provenances (i.e., a species from a particular location) is among the first major decisions to be made for reforestation. Whether native or introduced trees are used, key questions are: "where will it grow?" and "how well will it grow?".

This chapter describes methods that have been developed to assist in predicting where and how well particular trees will grow in different environments. Though the methods described have mainly been developed to assist the introduction of trees they also are used to assist evaluating land for agriculture, horticulture and conservation as well as the assessment of pest and disease risks. Four main methods are considered: climatic interpolation to estimate conditions for any location, climatic mapping programs to show where particular species will grow, simple simulation models to predict how well particular species will grow at individual sites and simulation mapping programs to predict tree growth at many sites. Opportunities for and obstacles to the further application of these methods are considered in the discussion section.

## CLIMATIC INTERPOLATION

Climate has an important influence on plant growth, and climatic analysis is useful as a means to predict where particular plants will grow. There is a long history in the use of climatic analysis methods to explain the natural distribution of plants and assist species introductions. The development of classification systems, such as those of Köppen (1918) and Thornthwaite (1948), was an important contribution. However, these relied on data from existing meteorological stations, so delineating the extent of each classification zone required a considerable amount of guesswork. It was not until the early 1980s that interpolation methods powerful enough to provide reliable estimates of mean climatic conditions for virtually any location were developed. Hutchinson et al. (1984) developed a set of mean monthly solar radiation relationships for Australia, which was one of the first interpolated climatic data sets to be developed at a continental scale to assist agroecological analysis. Since then, interpolation relationships for important mean monthly climatic factors such as maximum temperature, minimum temperature and precipitation have been developed for many major regions including Africa (Booth et al., 1989a), Latin America (Jones et al., 1990), and East/Southeast Asia (Zuo, 1997). Leemans and Cramer (1991) developed an interpolated global grid for temperature, precipitation and cloudiness. However, the meteorological stations they used were concentrated mainly in the USA and Europe, and the use of mean temperature rather than maximum and minimum temperatures was a limitation for agroecological studies. Monthly mean maximum temperature, minimum temperature and precipitation relationships also have been developed for individual countries including Australia (Hutchinson, 1995), Philippines (Jovanovic & Booth, 1996a), Indonesia (Jovanovic & Booth, 1996b), Zimbabwe (Booth et al., 1989b) and several others. Interpolated climatic data also have been developed for several regional studies (e.g., Nix et al., 1992, unpublished data; Hart et al., 1996).

Activity in developing interpolation relationships has increased rapidly in recent years with groups such as the Centre for Resource and Environmental Studies, Australian National University (Hutchinson et al., 1996), Potsdam Institute for Climate Impact Research (Cramer, 1996), International Irrigation Management Institute/Utah Climate Center (IIMI/Utah Climate Center, 1997) and Climate Research Unit, University of East Anglia all involved with developing interpolation relationships for individual countries, major regions or the whole world. Most centers are using the thin plate spline methods developed by Hutchinson (1995), though the IIMI/Utah Climate Center team uses an updated derivative of minimum curvature routines (D. Jensen, 1997, personal communication). The methods all generally relate climatic factors to latitude, longitude and elevation. Whichever method is used, results are typically within about ±0.8°C for temperature. For precipitation, errors are typically within ±10% for most locations, but may be much greater in some instances in mountainous or coastal regions (see, for example, Hutchinson et al., 1996; IIMI/Utah Climate Center, 1997).

Once a climatic interpolation relationship has been created, data can be estimated for grids of any size if a digital elevation model (i.e., a mathematical representation of topography) is available. Grids as coarse as half a degree (i.e., approx. 50 km) were used for some early global and continental studies (e.g., Booth, 1989a), while grids as fine as 200 m or less have been used for regional studies (e.g., Nix et al., 1992, unpublished data). The resolution chosen is usually a compromise between the advantages of greater detail and the disadvantages of storing, analyzing and displaying more information.

The ability to estimate mean climatic conditions reliably for any location has opened up great opportunities for biogeographic analysis. For example, the BIO-CLIM system devised by Nix, Busby, and Hutchinson (Hutchinson et al., 1984; Nix, 1986; Busby, 1991) takes in geocoded (i.e., latitude, longitude and elevation) data for a particular taxon, analyses the range of mean climatic conditions experienced, and generates a description of climatic requirements. One of the first applications of BIOCLIM was to analyze the natural distribution of *Eucalyptus citriodora* in Australia and to suggest suitable areas for growing the species in Africa (Booth, 1985). Though analyzing natural distributions can provide an initial indication of a species' climatic requirements, many species can grow well under conditions that are somewhat different from those where they grow naturally. Booth et al. (1988) used climatic interpolation relationships for Africa to analyze results from trial sites in Africa and develop improved descriptions of species requirements.

BIOCLIM also provides a method for identifying plantations of any species that may be particularly vulnerable to climatic change. For instance, present climatic conditions at 71 major plantation sites, which represent about 90% of the total area of *Pinus radiata* in Australia, were analyzed and a simple climate change scenario used to estimate possible conditions in 50 yr (Booth & McMurtrie, 1988). While sophisticated models cannot yet predict with any confidence the effects of climatic and atmospheric change on forests (Landsberg, 1996) the much simpler BIOCLIM analysis does at least identify plantations currently experiencing the most extreme conditions. These could be monitored to provide the earliest possible warning of possible impacts.

## CLIMATIC MAPPING

Climatic interpolation relationships are powerful tools for determining the climatic requirements of particular species by analyzing their natural distributions and also results from trials. However, it can be difficult to visualize where suitable areas exist if only a written description such as the following for the Petford provenance of *Eucalyptus camaldulensis* is available: mean annual rainfall, 400 to 2500 mm; rainfall seasonality, summer; dry season length, 2 to 8 mo; mean maximum temperature, hottest month, 28 to 38°C; mean minimum temperature, coldest month, 6 to 22°C; mean annual temperature, 18 to 28°C.

Climatic mapping programs were developed to provide a quick and easy means of visualizing areas that satisfied particular sets of climatic conditions. Figure 5–1 shows output from a climatic mapping program for China that uses interpolated climatic data estimated for about 100 000 locations in a 0.1° grid (approximately 10-km spacing). The map shows areas that satisfy the description of climatic requirements for the Petford provenance of *Eucalyptus camaldulensis* shown above.

In contrast to the BIOCLIM program, which was initially devised to study natural distributions, climatic mapping programs were devised to assist plant introductions. While various versions of the BIOCLIM program have used 12, 24 or 36 different climatic factors, climatic mapping programs have used a much smaller set of six or seven factors. Initially these were the six factors listed above, which were used by Webb et al. (1980, 1984) in their *Guide to Species Selection for Tropical and Sub-Tropical Plantations*. Using these factors meant that 175 descriptions of tree species requirements were immediately available from the Webb et al. (1984) publication.

Fig. 5–1. Output from a climatic mapping program for China (Yan Hong et al., 1996). Dark-shaded areas are climatically suitable for the Petford provenance of *Eucalyptus camaldulensis*. Similar mapping programs have been developed for several other countries and regions (see text).

The first climatic mapping programs were developed for Africa (Booth et al., 1989a) and Zimbabwe (Booth et al., 1989b). Subsequently, programs have been developed at CSIRO Forestry and Forest Products for several countries including Australia, China, Thailand, Indonesia, Vietnam, Laos, Cambodia and the Philippines (Booth, 1996a). At coarser scales, programs also have been developed for regions including Central America, South America, Southeast Asia and the whole world. The WORLD program was used to improve descriptions of the climatic requirements of 21 Australian tree species, including some of the commercially most important eucalypt species (Booth & Pryor, 1991).

Generally, the six factors listed at the beginning of this section have proved to be useful for discriminating areas climatically suitable for particular species. However, in areas affected by frosts, it has been found to be useful also to include absolute (i.e., record) minimum temperature. Some locations occasionally experience record minima that are significantly much cooler than those at other locations that have similar mean minima of the coldest month. For example, Australian eucalypt (*Eucalyptus*) and acacia (*Acacia*) species have been successfully grown in parts of Russia, China and the USA for several years only to suffer severe losses when extreme frosts are experienced (see, for example, Yan Minquan et al., 1991).

The inclusion of interpolated absolute minimum temperature helps to identify these vulnerable areas. For example, Booth and Yan Hong (1991) analyzed conditions at *Acacia mearnsii* plantations and determined that the species could not be recommended for planting at areas where the absolute minimum temperature fell below −5°C. Interpolation surfaces developed for China were used to estimate absolute minimum temperature conditions at several recently established *Acacia mearnsii* trial sites in China (Booth et al., 1994). For Longdouxie (24°57′ N lat, 115°5′ W long) the predicted absolute minimum temperature using interpolation was −7.1°C, though the lowest temperature recorded at the nearest meteorological station was −4.3°C. In the event, the most severe frost event for 50 yr resulted in an absolute minimum temperature of −7.6°C recorded at Longdouxie and the *Acacia mearnsii* trees were all killed (Yan Minquan et al., 1994). The same major frost event resulted in several hundred million dollars of damage to plantations and citrus orchards in southern China.

Another possible limitation of the Webb et al. (1984) factors is the use of rainfall of less than 40 mm to determine dry season months. In many tropical areas this can result in no dry season being identified even though plants may experience some water stress. A level of 100 mm may be more appropriate for use in some regions. Inclusion of information on rainfall variability (e.g., Samani & Hargreaves, 1989) also would be a useful improvement.

The use of climatic mapping programs is now well established and similar systems are incorporated into tree selection systems developed by other agencies. For example, the *Tree and Shrub Selection Guide* developed by the Environmentek/CSIR in South Africa (von Maltitz et al., 1996) includes bioclimatic indices for 200 species and a climatic mapping program for seven factors for a 1.6-km grid across the country.

## SIMPLE SIMULATION MODELS

The ability to identify where particular species will grow is useful, but most potential growers also want an indication of how well a particular plant is likely to grow. In agriculture an enormous effort has gone into developing sophisticated models to predict the potential yield of the dozen or so major crop plants, including rice (*Oryza sativa* L.), maize (*Zea mays* L.) and wheat (*Triticum aestivum* L.), which dominate world production (e.g., Richie, 1994; IBSNAT, 1994). However, there are hundreds of other species of lesser importance for agriculture, horticulture and forestry, including many plantation trees, for which such sophisticated models are not and will not be available in the foreseeable future. There is a need for simpler models, which can provide at least a general indication of the potential of selected sites for particular species.

The organization CSIRO developed a model called Plantgro to provide coarse predictions of the potential growth of lesser-known plants (Hackett, 1991). The model is run using individual plant, climate and soil files and the program makes it easy for users to develop new files to meet their own particular requirements. The program uses notional relationships that describe a particular plant's requirements for a particular factor in terms of suitability ratings ranging between 1 (ideal conditions) and 0 (highly unsuitable conditions). It includes water balance and non-linear heat sum calculations, but the suitability relationships for the 11 soil factors, such as salinity, N and P, are described by simple two-dimensional graphs. The effects of the individual factors are combined using Liebig's Law of the Minimum (Liebig, 1855). In simple terms this means that the most limiting factor in any one month determines the level of a plant's performance. Davidson (1996) has described how plant files can be developed to describe the requirements of particular tree species. Hackett (1997, personal communication) has developed 1700 preliminary Plantgro plant files, including files for about 371 tree species, using information from FAO's ECOCROP1 database (Diemer, 1996). The ECOCROP1 includes brief descriptions of species requirements in a standard format and is intended to assist identifying suitable crops for specific uses and/or soil and climatic environments around the world.

Over 500 copies of the Plantgro package (Hackett, 1991) have been sold, which indicates the significant demand for this type of package. It has been integrated with a complex geographical information system and used as the basic predictive engine for a multimillion dollar study of plantation forestry potential across 40% of Indonesia (NMFP, 1994; Pawitan, 1996). Though the consultants who completed this study expressed confidence in the system there are as yet few examples of published work validating the model. It is difficult to find large data sets for plantation tree species that provide both growth and detailed environmental data. Those which do exist often lack critical factors such as measures of available water holding capacity or are not available for model testing for commercial reasons. Fryer (1996) carried out a small test for nine sites of *Eucalyptus camaldulensis* growing in five Central American countries and obtained an $r^2$ of 0.63 between Plantgro's ratings and mean annual height increment. The Brazilian research agency EMBRAPA is currently evaluating Plantgro using information from its extensive records of eucalypt trials, so it is hoped that tests with larger data sets will be carried out soon (A. Higa, 1997, personal communication).

In the last 15 yr or so considerable effort has gone into developing complex process-based models to simulate tree growth (e.g., McMurtrie et al., 1990; Wang & Jarvis, 1990). While these have provided some useful research insights they have not been designed for routine site matching work. Two Australian groups have recently developed much simpler models with a view to ultimately providing programs suitable for use by forest managers. The work of the first group has been described by Battaglia and Sands (1997). They have developed a model called ProMod, which is based firmly on physiological knowledge, but can be run using the quantity and quality of data generally available to forest managers when evaluating prospective plantation sites. These data include latitude, longitude, elevation, slope and aspect of the site and a classification of soil depth, texture, stoniness, drainage and a rating of soil fertility. Mean climatic data for the site are estimated using the interpolation methods mentioned earlier. The model provides estimates of mean annual increment for particular species on particular sites. The model was initially developed using data from nine *Eucalyptus globulus* sites in southern Tasmania and Western Australia. It was then validated using independent data from 19 sites in northern Tasmania. Predictions of mean annual increment produced by the model were highly correlated with the actual values $r^2 = 0.81$) and were clearly superior to estimates produced by the existing site classification system ($r^2 = 0.15$) or multiple regression ($r^2 = 0.30$).

The work of the second group has been described by Landsberg and Waring (1997) who have developed a model called 3-PG (Physiological Principles in Predicting Growth) that calculates total C fixed (i.e., gross primary production) from utilizable, absorbed photosynthetically active radiation obtained by correcting the photosynthetically active radiation absorbed by the forest canopy for the effects of soil drought, atmospheric vapor pressure deficits and stand age. The model requires readily obtainable soil and weather data as input and runs on monthly time steps. The model has shown encouraging results relating predicted net primary production to measured wood production of *P. radiata* at eight sites in New Zealand and Australia ($r^2 = 0.82$) (Coops et al., 1998).

Plantgro, ProMod and 3-PG each have certain advantages and there may be a role for all three models in site evaluation. For instance, Plantgro can be calibrated using much less detailed information than either of the other two models and so is more suited to providing estimates of performance for lesser-known species. ProMod provides more reliable estimates of performance, but requires much more detailed information to be calibrated for a particular species. ProMod is suitable for evaluating the performance of comparatively intensively studied species such as *Eucalyptus grandis*, *E. globulus* and *P. radiata*, for which detailed physiological information is already available. The 3-PG also requires relatively detailed physiological information, but has an advantage over ProMod in predicting the likely growth of a stand over time and also has the potential to provide important information on stem size distributions.

## SIMULATION MAPPING PROGRAMS

Predicting growth for individual locations is useful, but often there is a need to predict growth across whole regions. Personal computer-based simulation mapping programs were developed so that a simplified version of the Plantgro program could be run for thousands of locations. They carry out the same climatic calculations as the full Plantgro model including assessments of water balance, nonlinear heat sum and solar radiation status. However, they use much simpler soil fertility estimates, which can be derived from maps. The programs generate three maps showing separate limitations for soil and climatic factors as well as a map showing their combined limitations. A moveable marker can be placed over any location and a summary analysis of month-by-month limitations for major factors such as solar radiation, water and temperature can be examined. Simulation mapping programs do not replace the need for detailed evaluations of individual sites, but they do provide a useful broad overview, which can assist in checking descriptions of species' requirements. The first simulation mapping program was developed for Africa (Booth, 1991) and subsequently programs have been developed for Australia, China and Thailand. The programs for Africa, Australia, China and Thailand include data for 10 187, 11 299, 15 789, and 6 242 locations respectively. Color outputs from all four programs have been presented by Booth (1996b).

## DISCUSSION

The development of climatic interpolation relationships for many areas of the world has provided a powerful tool for assisting species-site matching. The interpolation relationships have not only made it possible to analyze conditions within a species' natural distribution, but also to determine conditions at trial sites and to generate maps showing areas that are likely to be climatically suitable for that species.

Though reasonably reliable climatic interpolation results can be obtained for most parts of the world, errors can be unacceptably high where the coverage used is inadequate, particularly in mountainous and coastal regions. For example, interpolation estimates from the IIMI/Utah Climate Center (1997) World Water and Climatic Atlas indicate all locations within Kerala State in southern India have mean annual precipitations below 3000 mm, while actual data from 49 of 92 stations in the region record over 3000 mm, with the highest at 5880 mm (India Meteorol. Dep., 1962). The generalized cross validation methods included in some interpolation programs can help to identify areas with high error levels (see, for example, Booth et al., 1989b). In some cases it may be possible to obtain additional data from these areas. For example, data may be obtained from more meteorological stations or from farms that sometimes have good records, but may not report to the national meteorological service. Unfortunately, in many countries such records may be limited and even data collection from existing national networks of recording stations is often deteriorating due to cutbacks in government funding. Unless reliable means can be devised to estimate weather conditions from satellites, the deterioration in the quantity and quality of ground-based records will become a significant obstacle to developing improved climatic interpolation relationships in future years.

The widespread use of CD-ROM technology is helping to make large databases of interpolated climatic information readily available to potential users. For example, large databases for Africa (Hutchinson et al., 1996) and Asia (IIMI/Utah Climate Center, 1997) have been released in the last couple of years. Useful as these databases are, many potential users have neither the necessary skills nor access to sophisticated GIS tools to utilize the data fully. There is still a great need for simple low-cost systems, such as climatic mapping programs, which can be learned in a few minutes and used on a standard personal computer.

Just as the lack of basic information is a major impediment to developing improved climatic interpolations, so the lack of basic information from trials is a major limitation to developing improved descriptions of tree species' environmental requirements. The agricultural community realized the need for developing "minimum data sets" (MDS) for crops of major importance many years ago (Nix, 1984). The MDS were defined as the minimum amount of environmental, crop and management data a model user would need to validate and apply existing crop models. The concept provided the basis for a major international research program lasting for a decade (IBSNAT, 1993) and the development of models suitable for practical application (IBSNAT, 1994). The circumstances of the forestry community now are in some ways comparable to those of the agricultural community when the IB-SNAT project was established in about 1983. A number of models such as Plantgro, ProMod, 3-PG, as well as other more complex models, are ready for global testing. The field testing of these models across a wide range of sites would expose imperfections, allow the models to be refined and result in fully practical tools for site matching.

However, funding for large international research programs is much more difficult to obtain than it was 15 yr ago. So, in the absence of a well-coordinated international program it is necessary for individuals and organizations to make the most of more limited resources. For example, the Center for International Forestry Research (CIFOR) is funding a Tree Growth and Permanent Plot Information System (TROPIS) database (J. Vanclay, 1997, personal communication). Though this does not include detailed growth or environmental data, it does allow researchers to identify quickly where particular species and provenances have been included in trials and provides contact details for requesting more detailed information. The TROPIS provides the potential for researchers to track down information from individual trials, as well as from databases such as TREDAT (Brown et al., 1989) and MPTS (von Carlowitz, 1991), which contain detailed growth and environmental data.

As models such as Plantgro, ProMod and 3-PG are validated using information derived from TROPIS and other sources, it will be desirable to link them to environmental databases so that predictions can be generated for thousands of locations. The ability to generate predictions for thousands of locations is very important when validating models. For example, the Webb et al. (1980, 1984) compendia contain a description of the climatic requirements of *P. radiata*, which suggests the species would be unsuitable for growing in New Zealand. This is an obvious error in the species description, as there are in fact more than a million hectares of *P. radiata* plantations in New Zealand. The problem went undetected when the description was presented in the form of a written description, but was quickly identified when the description was run through a climatic mapping program. Using the mapping program, it also was easy to correct the description. Simulation mapping programs also can be used to check more complex models, though the lack of minimum data sets is a major problem for validating models across broad areas.

Another major impediment to the development of simulation mapping programs is the lack of reliable soil information. Broad-scale databases of soil information (e.g., FAO, 1996) are becoming available on CD-ROM in a similar way to climatic information, but the difficulties of interpolation are much greater for soil factors than climatic factors. While there are some links between soil morphogenesis and climatic and topographic conditions, there is a need for much improved methods of estimating important soil factors, such as water holding capacity and nutrient levels, across broad areas.

A further challenge is incorporating the assessment of pest and disease risks in models. For example, current work in Vietnam, Thailand, and Australia evaluates the use of some of the methods described here to assess the vulnerability of particular areas for *Cylindrocladium quinqueseptatum,* a leaf blight disease of *Eucalyptus camaldulensis.*

Great improvements have been made in matching genotypes to environment over the last 10 to 15 yr, particularly with the development of climatic interpolation and improved simulation models. Though analysis methods can prevent major mistakes and suggest appropriate species and provenances to test in particular areas they are as yet no substitute for actual trials. Trials should always be established before major plantations are established. It also is important to consider the socioeconomic impact of plantations on local populations as well as the long-term sustainability of proposed plantations (Nambiar & Brown, 1997).

# REFERENCES

Battaglia, M., and P. Sands. 1997. Modeling site productivity of *Eucalyptus globulus* in response to climatic and site factors. J. Plant Physiol. 24:831–850.

Booth, T.H. 1985. A new method for assisting species selection. Commonweal. For. Rev. 64:241–249.

Booth, T.H. 1991. A climatic/edaphic database and plant growth modeling system for Africa. Ecol. Model. 56:127–134.

Booth, T.H. (ed.). 1996a. Matching trees and sites. Proc. Int. Workshop, Proc. no. 63, Bangkok, Thailand. 27–30 Mar. 1995. ACIAR, Canberra, Australia.

Booth, T.H. 1996b. Simulation mapping programs for Africa, China, Thailand and Australia. p. 118–123. *In* T.H. Booth (ed.) Matching trees and sites. Proc. Int. Workshop, Proc. no. 63, Bangkok, Thailand. 27–30 Mar. 1995. ACIAR, Canberra, Australia.

Booth, T.H., T. Jovanovic, and Y. Hong 1994. Climatic analysis methods to assist choice of Australian species and provenances for frost-affected areas in China. p. 26–31. *In* A.G. Brown (ed.) Australian tree species research in China. Proc. no. 48, Zhangzhou, Fujian Province, People's Republic of China. 2–5 Nov. 1992. ACIAR, Canberra.

Booth, T.H., and R.E. McMurtrie. 1988. Climatic change and *Pinus radiata* plantations in Australia. p. 534–545. *In* G.I. Pearman (ed.) Greenhouse : Planning for climate change. CSIRO, Melbourne.

Booth, T.H., H.A. Nix, M.F. Hutchinson, and T. Jovanovic. 1988. Niche analysis and tree species introduction. For. Ecol. Manage. 23:47–59.

Booth, T.H., and L.D. Pryor. 1991. Climatic requirements of some commercially important eucalypt species. For. Ecol. Manage. 43:47–60.

Booth, T.H., J.A. Stein, H.A. Nix, and M.F. Hutchinson. 1989a. Mapping regions climatically suitable for particular species: an example using Africa. For. Ecol. Manage. 28:19–31.

Booth, T.H., J.A. Stein, M.F. Hutchinson, and H.A. Nix. 1989b. Identifying areas within country climatically suitable for particular tree species: An example using Zimbabwe. Int. Tree Crops J. 6:1–16.

Booth, T.H., and Y. Hong 1991. Identifying climatic areas in China suitable for *Acacia mearnsii* and *Acacia mangium*. p 52–56. *In* J.W. Turnbull (ed.) Proc. Advan. Trop. Acacia Res., Proc. no. 35, Bangkok, Thailand. 11–15 February. ACIAR, Canberra, Australia.

Brown, A.G., L.J. Wolf, P.A. Ryan, and P. Voller. 1989. TREDAT: A tree crop database. Aust. For. 52:23–29.

Busby, J.R. 1991. BIOCLIM–a bioclimate prediction system. p. 64–68. *In* C.R. Margules and M.P. Austin (ed.) Nature conservation—cost effective biological surveys and data analysis. CSIRO, Melbourne, Australia.

Coops, N.C., R.H. Waring, and J.J. Landsberg. 1998. Assessing forest productivity in Australia and New Zealand using a physiologically-based model driven with averaged monthly weather data and satellite derived estimates of canopy photosynthetic capacity. For. Ecol. Manage. 104:113–127.

Cramer, W. 1996. Data requirements for global terrestrial ecosystem modeling. p. 30–37. *In* B. Walker and W. Steffen (ed.) Global change and terrestrial ecosystems. Cambridge Univ. Press, Cambridge, MA.

Davidson, J. 1996. Developing Plantgro plant files for forest trees. p. 93–96. *In* T.H. Booth (ed.) Matching trees and sites. Proc. Int. Workshop, Proc. no. 63, Bangkok, Thailand. 27–30 Mar. 1995. ACIAR, Canberra, Australia.

Diemer, P. 1996. ECOCROP1 DOS Version 1.1. Soil Resour., Manage. Conserv. Serv., FAO, Rome.

Food and Agriculture Organization. 1996. Digital soil map of the world and derived soil properties. CD-ROM Vers. 3.5. FAO, Rome.

Fryer, J. 1996. Climatic mapping for eucalypts in Central America. p. 50–55. *In* T.H. Booth (ed.). Matching trees and sites. Proc. Int. Workshop, Proc. no. 63, Bangkok, Thailand. 27–30 Mar. 1995. ACIAR, Canberra, Australia.

Hart, T.C, S.L. Greene, and A. Afonin. 1996. Mapping for germplasm collections: Site selection and attribution. p. 111–122. *In* Proc. 3rd Int. Conf./Workshop Integrating GIS and Environ. Model., Santa Fe, TX. 21–26 January. Natl. Center Geogr. Inform. Anal., Santa Barbara, CA.

Hackett, C. 1991. Plantgro: A software package for coarse prediction of plant growth. CSIRO, Melbourne, Australia

Hong, Y., T.H. Booth, and H. Zuo 1996. GREEN—a climatic mapping program for China and its use in forestry. p. 24–29. *In* T.H. Booth (ed.) Matching trees and sites. Proc. Int. Workshop, Proc. no. 63, Bangkok, Thailand. 27–30 Mar. 1995. ACIAR, Canberra, Australia.

Hutchinson, M.F. 1995. Interpolating mean rainfall using thin plate smoothing splines. Int. J. GIS 9:385–403.

Hutchinson, M.F., T.H. Booth, J.P. McMahon, and H.A. Nix. 1984. Estimating monthly mean values of daily solar radiation for Australia. Solar Energy 32:277–290.

Hutchinson, M.F., H.A. Nix, J.P. McMahon, and K.D. Ord. 1996. The development of a topographic and climate database for Africa. p. 50–64. Proc. 3rd Int. Conf./Workshop on Integrating GIS and Environ. Model., Santa Fe, NM. 21–25 January. Natl. Center Geogr. Inform. Anal., Santa Barbara., CA.

International Benchmarks Sites Network for Agrotechnology Transfer. 1993. International Benchmarks Sites Network for Agrotechnology Transfer, the IBSNAT decade. Dep. Agron. Soil Sci., College of Trop. Agric. Human Resour., Univ. Hawaii, Honolulu.

International Benchmarks Sites Network for Agrotechnology Transfer. 1994. The decision support system for agrotechnology transfer. Vers. 3.0, User's Guide. Dep. Agron. Soil Sci., College of Trop. Agric. Human Resour., Univ. Hawaii, Honolulu.

International Irrigation Management Institute/Utah Climate Center. 1997. World water and climate atlas—Asia. CD-ROM. IIMI/Utah Climate Center, Utah State Univ., Salt Lake City, UT.

India Meteorological Department. 1962. Monthly and annual normals of rainfall and of rainy days. Part 3. Memoirs Dep. Vol. 31. Gov. India Press, New Delhi.

Jones, P.G., D.M. Robinson, and S.E. Carter. 1990. A geographical information approach for stratifying tropical Latin America to identify research problems and opportunities in natural resource management for sustainable agriculture in CIAT. Agroecolog. Stud. Unit, CIAT, Cali, Colombia.

Jovanovic, T., and T.H. Booth. 1996a. The development of climatic interpolation realtionships for the Philippines. p. 56–64. In T.H. Booth (ed.) Matching trees and sites. Proc. Int. Workshop, Proc. no. 63, Bangkok, Thailand. 27–30 Mar. 1995. ACIAR, Canberra, Australia.

Jovanovic, T., and T.H. Booth. 1996b. The development of interpolated temperature and precipitation relationships for the Indonesian Archipelago. p. 30–37. In T.H. Booth (ed.) Matching trees and sites. Proc. Int. Workshop, Proc. no. 63, Bangkok, Thailand. 27–30 Mar. 1995. ACIAR, Canberra, Australia.

Köppen, W. 1918. Klassifikation der Klimate nach Temperatur, Niederschlag und Jahreslauf. Petermanns Geographische Mitteilungen, 64:193–203.

Landsberg, J.J. 1996. Impact of climate change and atmospheric carbon dioxide concentration on the growth of planted forests. p. 205–221. In W. Bousma et al. (ed.) Greenhouse: Coping with climate change. CSIRO, Melbourne, Australia.

Landsberg, J.J., and R.H. Waring. 1997. A generalised model of forest productivity using simplified concepts of radiation-use efficiency, carbon balance and partitioning. For. Ecol. Manage. 95:209–228.

Leemans, R., and W.P. Cramer. 1991. The IIASA database for mean monthly values of temperature, precipitation and cloudiness on a global terrestrial grid. Publ. RR-91-18. Inst. Appl. Syst. Anal., Laxenburg, Austria.

Liebig, J. 1855. Die Grundsätze der Agriculturchemie mit Rücksicht auf die in England angestellten Untersuchungen. F. Vieweg und Sohn, Braunschweig.

McMurtrie, R.E., D.A. Rook, and F.M. Kelliher. 1990. Modelling the yield of Pinus radiata on a site limited by water and nitrogen. For. Ecol. Manage. 30:381–413.

Nambiar, E.K.S., and A.G. Brown (ed.). 1997. Management of soil, nutrients and water in tropical plantations. CSIRO/ACIAR/CIFOR, Canberra, Australia.

Minquan, Y., Z. Yutian, Z. Xijin, and Z. Xiangsheng 1994. Effect of low temperatures on Acacia. p. 176–179. In A.G. Brown (ed.) Australian tree species research in China. Proc. Int. Workshop, Proc. no. 48, Zhangzhou, Fujian Province, People's Republic of China. 2–5 Nov. 1992. ACIAR, Canberra, Australia.

Nix, H.A. 1984. Minimum datasets for agrotechnology transfer. p. 181–188. In V. Krumble (ed.) Proc. Int. Symp. Minimum Datasets for Agrotechnol. Transfer, Patancheru, India. 10–11 June 1983. ICRISAT, India.

Nix, H.A. 1986. A biogeographic analysis of Australian elapid snakes. p. 4–15. In R. Longmore (ed.) Atlas of Australian Elapid Snakes. ACT, Bur. Flora and Fauna, Canberra, Australia.

National Masterplan for Forest Plantations. 1994. The National Masterplan for Forest Plantations Report. The Ministry Forestry Rep. Indonesia/DHV Consult., PT Tricon Jaya and PT Catur Tunggal Sarana Consultants, Jakarta.

Pawitan, H. 1996. The use of Plantgro in forest plantation planning in Indonesia. p. 97–100. In T.H. Booth (ed.) Matching trees and sites. Proc. Int. Workshop, Proc. no. 63, Bangkok, Thailand. 27–30 Mar. 1995. ACIAR, Canberra, Australia.

Ritchie, J.T. 1994. Classification of crop simulation models. p. 3–14. In P.F. Uhlir and G.C. Carter (ed.) Crop modelling and environmental data. CODATA Monogr. Ser. Vol. 1. CODATA, Paris.

Samani, Z.A. and G.H Hargreaves. 1989. Applications of a climatic database for Africa. Comput. Electron. Agric. 3:317–325.

Thornthwaite, C.W. 1948. An approach towards a rational classification of climate. Geograph. Rev. 38:55–94.

Von Carlowitz, P.G., G. Wolf, and R. Klemperman. 1991. Multipurpose tree and shrub Database Vers. 1.0. User's manual. ICRAF, Nairobi, Kenya.

Von Maltitz, G., P. Brown, and A. Tapson. 1996. Tree and shrub selection guide. Environmentek, CSIR, Pretoria.

Wang, Y.P. and P.G. Jarvis. 1990. Description and validation of an array model—MAESTRO. Agric. For. Meteorol. 51:257–280.

Webb, D.B., P.J. Wood, and J.P. Smith. 1980. A guide to species selection for tropical and sub-tropical plantations. Commonweal. For. Inst., Oxford, United Kingdom.

Webb, D.B., P.J. Wood, J.P. Smith, and G.S. Henman. 1984. A guide to species selection for tropical and sub-tropical plantations. 2nd ed.. Commonweal. For. Inst., Oxford, United Kingdom.

Zuo, H. 1997. Agroclimatic analysis for mainland East Asia by a GIS approach. Ph.D. diss. Australian Natl. Univ.

# 6    Germplasm Collecting Using Modern Geographic Information Technologies: Directions Explored by the N.I. Vavilov Institute of Plant Industry

**Alexandr Afonin**

*N.I. Vavilov Institute of Plant Industry*
*St. Petersburg, Russia*

**Stephanie L. Greene**

*Washington State University*
*Prosser, Washington*

The largest gene bank in the world, indeed the global system as a whole, cannot hold and maintain all the genetic variation present in cultivated plants and their wild relatives. Recognizing they cannot preserve all genetic variability, most gene banks have adopted a triage approach for ex situ conservation. Accessions are added to ex situ collections but at the expense of conserving other germplasm. From this perspective, the question "what do we take into an ex situ collection?" acquires a dramatic subtext. Genetic resource specialists at the N.I. Vavilov Institute of Plant Industry (VIR) use several criteria to determine if an accession should be added to a collection: (i) Has the accession a specific agronomic trait that is missing from the current collection? (ii) Is the accession genetically different from other samples in the collection? (iii) Has the population from which the accession has been sampled threatened with disappearance? The criteria reflect the efforts to develop ex situ germplasm collections that represent the genetic diversity in a crop gene pool, with an emphasis on germplasm that is potentially useful and/or in need of protection.

The VIR Department of Plant Introduction is exploring the use of modern Geographical Information System (GIS) technologies to increase the efficiency of adding new acquisitions to existing germplasm collections. Three groups of thematic data are considered particularly useful: (i) the geographic distribution of specific environmental factors that limit the occurrence of crops and wild relatives, (ii) the distribution of wild forms and related species and regional distribution of cultivars, (iii) the distribution of collection sites where germplasm has been previously collected and recommended areas for cultivars. As other chapters in this publication have suggested, enumerative manipulation of these types of datasets using GIS software provides opportunities to optimize the content of existing collections and carry out goal-directed plant collection and introduction. For example, combining Datasets 1 and 2 allows us to identify areas with unfavorable environmental factors (UEF) within areas of cultivation or occurrence. This provides us with direction for collecting accessions tolerant to the UEF. Combining Datasets 2 and 3 allows us to identify gaps and redundancy in existing collections. A number of practical applications can be developed once the initial datasets are developed. The purpose of this chapter is to provide some concrete examples to illustrate the approaches explored at VIR to use GIS technologies to guide the development of worldwide collections of plant germplasm. Our examples make use of the GIS database developed by Hart at al. (1996) to collect forage species in the North Caucasus Mountains, Russia. We show how these cartographic datasets can be coupled with on-site population data to predict the intraregional distribution of species and guide the acquisition of germplasm tolerant to unfavorable environments. We then illustrate how remotely sensed and secondarily derived geographic information can be used to estimate sampling frequencies and assess level of anthropogenic disturbance. We conclude the chapter by discussing efforts by VIR to expand these techniques to guide acquisition of germplasm on a global scale.

## WHERE TO SEARCH FOR WILD SPECIES

Wild or naturalized forms of cultivated species and their relatives are frequently the targets of germplasm collection trips. Although they have distinct distribution patterns in nature, reflecting the ecogeographic constraints of the species, these patterns may not be represented in the collections of gene banks, unless special efforts have been made to ensure thorough representation. Comparing maps illustrating where ex situ accessions have been sampled with species distribution maps, often reveals the fragmentation and unevenness of ex situ collections. One reason is that the plant hunter is frequently compelled to collect in close proximity to roads and trails, since undirected side departures take time and further resources. As a result the distribution of collection sites reflect road networks, instead of the real distribution of genetic variation in the territory. Although roadsides protected from grazing can be valuable collection sites for wild species, for locally cultivated species, roadways serve as a dispersal route for seed. As a result, if the objectives are to collect wild forms of cultivated species in areas where both forms are present, sampling along roadsides will probably be biased toward the cultivated gene pool. To avoid this, it is advisable to know where natural populations exist. Although frequently collectors can count on the help of local experts and herbarium information to target sites, this is not always the case. The possibilities of using GIS technologies to develop regional-scale prognostic maps of plant species distribution have been demonstrated by Afonin et al. (1997). The approach is somewhat similar to the method used to introduce valuable wood species to new regions (Booth, 1999, see Chapter 5; Booth, 1989). The assumption behind the approach is that the distribution of a species is defined by the environmental factors that limit growth. Modern methods of computer mapping allow us to create maps of environmental factors in a targeted region, and conduct a series of operative manipulations based on knowledge of the ecological amplitude of the targeted species. Using empirical data based on previous sampling data, we can calculate the probability of finding a species in a given area. With respect to each environmental factor, we can select territory that is "ecologically suitable" for the species in question. Such maps increase the efficiency of collecting trips since routes can be developed that exclude territory not having the appropriate environment to support the target species. In addition, these types of maps provide a way to quantify how optimal the habitat is, e.g., does the site represent a central or marginal position in the ecological amplitude of the species? This allows us to assess the competitiveness of the species and identify opportunities for reintroduction.

To demonstrate this approach, we use the geographic information and species data acquired from the 1995 joint Russian-American expedition to collect wild forage legumes and grasses in the northwest Caucasus Mountains, Russia. Greene et al. (1999a) provide a detailed description of the development of the GIS database and maps that classified agroclimatic conditions within a target region based on the most important limiting environmental factors. The main abiotic factors limiting the growth of perennial herbs in the north Caucasus Mountains are the climate (in particular minimum winter temperature), length of growing season, and territory aridity based on rainfall and temperature (Shiffers, 1953).

Germplasm was sampled from 149 sites distributed west to east from Taman to Nalchik (about 600 km) and from the north to south, from Krasnodar to Adler (more than 200 km). During the expedition, the presence or absence of wild populations of *Dactylis glomerata* L., *Galega orientalis* Lam., *Lotus corniculatus* L., *Medicago sativa* L. subsp. *falcata* (L.), *Trifolium fragiferum* L., *T. hybridum* L., *T. medium* L., *T. pratense* L., and *T. repens* L. was noted for each site visited. The GIS-derived climatic data of Greene et al. (1999a), combined with species occurrence data allowed us to estimate the ecological amplitude of each of the targeted species. For example, *M. sativa* subsp. *falcata* L. was encountered at elevations ranging from 0 to 2129 m, in areas where annual rainfall ranged from 466 to 1751 mm. *Galega orientalis* populations occurred in a more narrow range of elevation (303–1825 m) and precipitation (696–1529 mm). Defining the ecological amplitude of species based on empirical data allowed us to develop predictive maps for the region that will be useful in subsequent visits to the region. The maps were created using the GIS software, IDRISI for Windows 1.0 (The IDRISI Project, Clark Univ., MA).

To develop a prognostic map for the regional distribution of *G. orientalis* we queried the annual precipitation and elevation (Fig. 6–1a,b) maps to make new maps that identified the map grid cells that fell within the precipitation and elevation range where we had encountered *G. orientalis*, in the 1995 exploration. These two maps were then overlaid to produce a map indicating ecologically suitable territory for the species (Fig. 6–1c). It is interesting to note that the suitable territory for *G. orientalis* appears in several separate areas. We could speculate that genetic exchange between *G. orientalis* populations growing in separate areas would be limited, and that as a consequence we might expect to find greater genetic divergence between populations growing in the different areas, than between populations growing

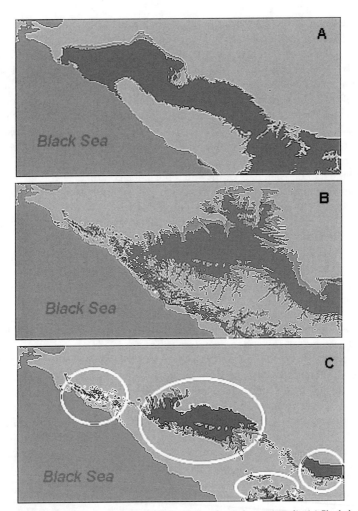

Fig. 6–1. Example of a regional distribution prognostic map for *Galega orientalis*. (*a*) Shaded area represents precipitation range corresponding to sites where *G. orientalis* was encountered. (*b*) Shaded area represents altitude range corresponding to sites where *G. orientalis* was encountered. (*c*) Shaded area predicts territory suitable for *G. orientalis*, in terms of precipitation and elevation.

within one of the contiguous areas, due to restricted gene flow (Greene & Hart, 1999, see Chapter 2) Although at this point, still a supposition, this example with *G. orientalis* illustrates the usefulness of using spatial analysis to identify areas where species are likely to occur, and to identify instances where populations within a species may be genetically divergent due to the discontinuities of habitat patches.

We could develop a more precise predictive model. To continue with our example, we know that meadow herbs such as *G. orientalis* do not grow under the canopy of forest trees. Using information from satellite imagery, we removed all the wooded areas from the ecologically suitable territory identified in the previous

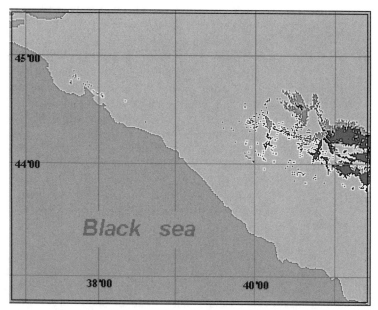

Fig. 6–2. Shaded area predicting ecologically suitable area for the occurrence of *Galega orientalis* in terms of precipitation, elevation, and forest coverage.

map to develop a more precise map indicating where *G. orientalis* might be collected (Fig. 6–2). We also can estimate the probability of encountering the species in the forecasted area using the ratio of the number of sites where *G. orientalis* was encountered to the total number of sites visited in the forecasted area (in 1995). Although we cannot say with certainty that the sites visited represent a random sampling of all sites, for the purposes of collecting germplasm, this estimated probability provides guidance when no other information is available, which is frequently the case. In our example, *G. orientalis* was encountered in 17 of the 67 sites visited in the forecasted area. Therefore the expected probability of encountering the species within the supposed borders is 0.25 (17 sites encountered divided by 67 total sites visited). We can evaluate the validity of our prognostic distribution maps if we test the model using site coordinates that were not included in the original calculations of the ecological amplitude. An error term can be derived based on the number of check sites that lay outside the forecasted distribution area. Although check sites may not be random, the existence of any previous information of species occurrence will be useful to the germplasm collectors in validating the prognostic distribution maps. Check sites *of G. orientalis* were obtained from herbarium specimens and expedition reports of past collection trips made by VIR. Of the 20 check sites identified, two did not lay within the forecasted area. Therefore the "validity" of the prognostic distribution map of *G. orientalis* in the territory of the northwest Caucasus was 90% although we would expect that the species was not a commonly occurring taxa since we encountered it in only 25% of the sites visited in the forecasted area.

# SEARCHING FOR GERMPLASM ADAPTED TO LESS
## THAN OPTIMAL ENVIRONMENTS

The relationship between place of origin and crop adaptation is rarely straight-forward, yet is the most common strategy for guiding the introduction of a crop into areas where it has not been previously grown. In a paper originally published in 1932, Vavilov (1992) cautioned that "climatic analogy" may be insufficient to predict crop adaptation. However, many studies of different plant species have demonstrated the relationship between tolerant ecotypes and the occurrence of unfavorable edaphic factors in the place of origin of the germplasm. For instance, Ab-Shukor et al. (1988) noted that the most salinity-tolerant populations of *T. repens* L. grew along the shores of a salt marsh. Salt tolerance was positively associated with geography in *M. sativa* subsp. sativa (Rumbaugh & Pendery, 1990). Increased Al tolerance was noted for a variety of cultivated plants growing in regions with acid soils. Beebe et al. (1997) were able to associate P efficiency in common bean (*Phaseolus vulgaris* L.) to geographic origin. Blum (1988) concluded that varieties, landraces or edaphic ecotypes of native populations that have been developed or evolved with the selection pressure of edaphic stress may constitute a useful source of resistance.

The relationship between climate of origin and crop adaptation is less clear. Hoffman and Parsons (1991) conclude that genetic variation exists within many species reflecting environmental adaptation to temperature and light intensity. In Russia, much work has been done connecting winter hardiness to climate of origin (e.g., Maximov, 1952; Tumanov, 1940). Now that we are able to carry out global-scale climate interpolations, there is much interest in developing climatic models that can be used to broadly predict species distribution and forecast the successful introduction of plant species (e.g., Booth et al., 1989; Booth, 1990; Jones et al., 1997). This same strategy may be effective in identifying potential collection areas with specific climatic regimes of interest to plant breeders. For example, a close relationship was found between frost tolerance and level of frost in a region of Russia where *T. pratense* is cultivated (Afonin & Chapurin, 1992).

Capitalizing on the link between adaptation and environmental conditions, geographic information and GIS analysis may therefore be an effective tool to aid in the search for tolerant germplasm. For instance, plant distribution maps can be combined with maps that identify areas with the environmental attributes of interest, allowing selection of regions that have a greater probability of yielding tolerant accessions, (if tolerance exists). Figure 6–3 shows an example of this procedure using data from the Caucasus Mountains. The following question was posed, "Where would it be best to sample *Trifolium ambiguum* in the North Caucasus mountains that exhibited drought tolerance?" A species distribution map (Fig. 6–3a) was built using the methods previously described for *G. orientalis*. Based on our 1995 observations, *T. ambiguum* was found in areas with an elevation range from 380 to 2485 m, and with an annual precipitation range of 516 to 2044 mm. Annual precipitation was selected as the best environmental attribute to characterized aridity (Fig. 6–3b). Combining the two maps we can see that the Taman peninsula and areas southeast of Stavropol may yield accessions best adapted to more arid conditions (Fig. 6–3c).

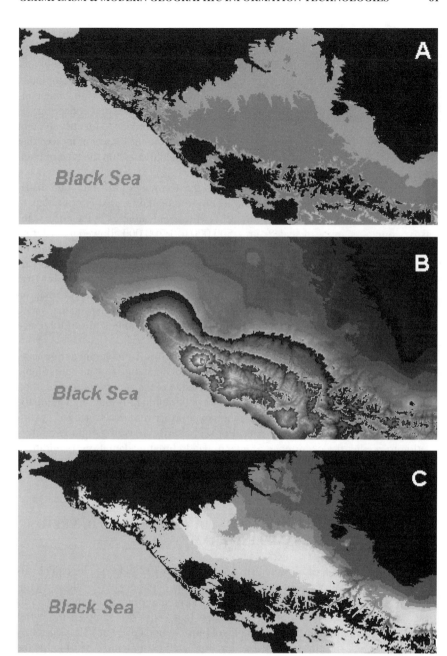

▉ 450 500...    ▦ 601 650...    ▦ 751 850...    ▦ >1200mm

Fig. 6–3. Prognostic map indicating distribution of *Trifolium ambiguum* adapted to arid conditions. (*a*) Dark grey and grey area represents area where *T. ambiguum* was encountered. Eighty percent of sites were located in dark grey areas. *Trifolium ambiguum* was not encountered in black areas. (*b*) Distribution of annual precipitation, and (*c*) areas shaded dark grey and grey may yield *T. ambiguum* germplasm adapted to arid conditions.

The use of limiting factor maps to search for tolerant germplasm needs to take into account that limiting factors can be defined by several agrometeorological indices. For instance, aridity can be quantified by annual precipitation, monthly precipitation, water deficit in soils, or by indices such as the hydrothermal factor index proposed by Selyaninov (Gringof et al., 1987, p. 18). A key point is that some indices are likely to be more effective for different plant species and collecting objectives then other indices. Using an ecogeographical approach to identify tolerant material must be preceded by careful work to determine the choice of indices that best reflects the influence of the targeted environmental stress on the type of plant under study.

It also is important to take into account the accuracy of the maps and distribution pattern, in particular, of the limiting environmental factors. For instance, in using soil maps to guide the collection of acid-tolerant samples in the north Caucasus Mountains, maps of scale from 1:500 000 to 1:200 000 allowed us to define regions with a prevalence of acidic soils. However, the soil pH measurements on site showed significant variation. For example, there were sites where soil pH varied from 4 to 7 along a 10-m transect. In such situations, the best approach is to use the small-scale maps as a general guide, but also to collect local site information on soil pH. A postcollection analysis can be carried out by upgrading soil maps with local site data in order to identify accessions that may exhibit acid soil tolerance (Greene et al., 1999b).

To effectively use an ecogeographic approach to collect tolerant germplasm, researchers must carefully select the best indices for characterizing an unfavorable environmental factor and have sufficient knowledge of the geographic distribution of the unfavorable factor. The use of GIS and geographic information to guide the collection of tolerant material is in its infancy. Further research is needed to confirm the usefulness of remotely sensed or secondarily derived geographic information to predict the occurrence of edaphic and climatic adaptation. In addition, care needs to be taken during collection to ensure populations reflecting adaptation are sampled (Greene & Hart, 1999, see Chapter 2).

## USING GEOGRAPHIC INFORMATION SYSTEMS TO DEVELOP A SAMPLING FRAMEWORK

The distribution of intraspecific genetic diversity of wild species can be difficult to predict. In absence of exact data, VIR plant explorers have followed a traditional approach based on uniform sampling at predetermined distance intervals based on the heterogeneity of the landscape (e.g., sites are sampled more frequently in mountains than plains). The Plant Introduction Office at VIR is currently exploring the use of remote-sensed and interpolated geographic data to guide the frequency of sampling based on habitat variability. This could promote more efficient sampling by targeting populations that reflect geographic differentiation. The ecogeographical approach to collecting wild species for germplasm collection means sampling populations throughout the species territory, but with more frequent collecting in areas with more ecological variation and in areas that represent transitions between ecogeographic zones. This means developing procedures that can quantify the degree of environmental diversity for different regions. Such a sampling framework would provide guidance in terms of recommending an optimal collection frequency for each territory. Below, we provide an example of how the Caucasus Mountain data sets could be used to develop a quantitative estimation of an Environmental Factors Diversity (EFD) index.

To begin, the region of interest was divided into a grid. Each grid cell in our example corresponded to one square degree (100 by 100 km) (Fig. 6–4). Temperature and precipitation are important environmental factors that influence the growth and distribution of plants. In this example we developed an EFD index using those data layers. We used IDRISI to estimate the range of temperature and precipitation within each grid cell. As an example, we can examine Cells 6 and 17. Precipitation ranges from 466 to 489 mm across the area represented by Grid 6. The difference between maximum and minimum is 23 mm. For Grid 17, precipitation ranges from 573 to 2025 mm, a difference of 1452 mm. The range of maximum air temperature during the growing period also is much greater in Cell 17, compared with Cell 6 (19°C vs. 2°C). To develop an index, we can standardize the maximum and minimal values in each of the 18 cells by expressing the difference between maximum and minimum as a percentage of the maximum range (e.g., maximum − minimum/maximum × 100 = %). The sum of the percentage values of each environmental factor of interest can then be used as an integral index of the environmental diversity in each cell. For example, in Cell 6, the EFD index estimated using annual precipitation and absolute maximum temperature during the growing season is 2 + 7 = 9%, compared with 100 + 70 = 170% for Cell 17. This would suggest that the environment in Cell 17 is much more diverse then the environment in Cell 6. Based on the index, frequency of sampling in Cell 17, which shows much more environmental diversity, may need to be sampled at a frequency at least an order of magnitude greater then the frequency of sampling in Cell 6, where the environment is much less diverse. We could use this approach for the quantitative estimation of environmental diversity in any territory and for any group of factors. Many GIS software packages include algorithms that can be used to carry out similar assessments of variability and pattern analysis.

## USING GEOGRAPHIC INFORMATION SYSTEMS TO ASSESS ANTHROPOGENIC INFLUENCE

In the instance when the distribution of wild crop species and cultivated germplasm of the same species is sympatric, as is the case of *M. sativa* subsp. *fal-*

Fig. 6–4. Example of environmental diversity estimation using geographical coordinates grid (see text).

*cata* and *T. pratense* in the Caucasus Mountains, care must be taken to avoid sampling cultivated escapes, if collection objectives are to sample wild populations. In estimating the genetic vulnerability of wild species, it also is important to assess the level of anthropogenic disturbance. Remote sensing can be used to objectively assess levels of anthropogenic disturbance. In the North Caucasus mountains, anthropogenic areas such as fields, cities and settlements are easy to identify on digital images. The ratio of anthropogenic areas to the whole area of a region could be used as a simple quantitative index to estimate the vulnerability of regional wild flora. Aside from obtaining a general indication of anthropogenic influence across the targeted area, we also can estimate and understand the dynamics of the anthropogenic processes occurring in each grid cell.

It would be particularly valuable to understand how patterns of disturbance are changing through time. This could be done by comparing disturbance levels based on satellite images taken 10 to 20 yr ago with more current satellite imagery. In addition, using remote sensed information we can understand what parts of the landscape are disturbed. For instance, the mountainous area of the Caucasus is on the whole characterized by a low level of anthropogenic influences (except for grazing and harvest of native hay meadows). However, from satellite imagery, it is evident that river and mountain valleys are used for fields and gardens. Different landscapes and their associated natural plant communities can be influenced by anthropogenic pressure even if the territory as a whole is not under pressure. In terms of developing collecting priorities, collection in areas with dynamic anthropogenic influences would be a priority.

## FUTURE DIRECTIONS

The examples we have discussed in this chapter illustrate several approaches currently being explored by the Office of Plant Introduction, at the Vavilov Institute, to guide the acquisition of plant germplasm at regional scales. The experience of developing and using the regional-scale cartographic data sets of the North Caucasus Mountains has stimulated interest in developing global-scale data sets that can be used to guide the acquisition of worldwide genetic resource collections. At VIR, a new department in the Plant Introduction Office is exploring the development of global-scale datasets. Ultimately, researchers envision the development of global maps that could be used to guide the search for germplasm adapted to the Russian climate. Considering the agricultural challenges we face in the next century, this work is essential. We would hope that publications such as this will attract the attention of, and motivate further collaboration among specialists to develop standardized global-scale GIS datasets appropriate for genetic resources conservation and use, and that these data sets are made freely and widely available (and in a format that lends them to many GIS platforms). Such data sets would be a valuable resource to the global plant germplasm community to guide the systematic acquisition of vulnerable germplasm and germplasm adapted to less than optimal environments.

# REFERENCES

Ab-Shukor, N.A., Q.O.N. Kay, D.P. Stevens, and D.O.F. Skibinski. 1988. Salt tolerance in natural populations of *Trifolium repens* L. New Phytol. 109:483–490.

Afonin, A.N., and V.F. Chapurin 1992. Winter tolerance of red clover in connection with origin [In Russian.] VIR Bull. 224:40–44.

Afonin, A.N., S.L. Greene, T. Hart, and J.J. Steiner. 1997. Use of ecological and geographical databases for prognostic intraregional plant distribution mapping. p. 6–8. *In* D. Geltman and Yu. Roskov (ed). Computer databases in botanical research. (In Russian.) Komorov Bot. Inst., St. Petersburg, Russia.

Beebe, S., J. Lynch, N. Galwey, J. Tohme, and I. Ochoa. 1997. A geographic approach to identify phosphorus-efficient genotypes among landraces and wild ancestors of common bean. Euphytica 95:325–336.

Blum, A. 1988. Plant breeding for stress environments. CRC Press, Boca Raton, FL.

Booth, T.H., J.A. Stein, H.A. Nix, and M.F. Hutchinson. 1989. Mapping regions climatically suitable for particular species: An example using Africa. Forest Ecol. Manage. 28:19–31.

Booth, T.H. 1990. Mapping regions climatically suitable for particular tree species at the global scale. Forest Ecol. Manage. 36:47–60.

Booth, T.H. 1999. Matching germplasm to geography: Environmental analysis for plant introduction. p. 63–74. *In* S.L. Greene and L. Guarino (ed.) Linking genetic resources and geography: Emerging strategies for conserving and using crop biodiversity. CSSA Spec. Publ. 27. ASA and CSSA, Madison, WI.

Greene, S.L., and T.C. Hart. 1999. Implementing geographic analysis in germplasm conservation. p. 25–38. *In* S.L. Greene and L. Guarino (ed.) Linking genetic resources and geography: Emerging strategies for conserving and using crop biodiversity. CSSA Spec. Publ. 27. ASA and CSSA, Madison, WI.

Greene, S.L., T.C. Hart, and A. Afonin. 1999a. Using geographic information to acquire wild crop germplasm: I. Map development and field use. Crop Sci. 39:836–842.

Greene, S.L., T.C. Hart, and A. Afonin. 1999b. Using geographic information to acquire wild crop germplasm: II. Post collection analysis. Crop Sci. 39:843–849.

Gringof, I.F., V.V. Popova, V.N. Strashnyy. 1987. Agrometeorology. (In Russian.) Gydrometeoizat Press, Leningrad, USSR.

Hart, T.C., S.L. Greene, and A. Afonin. 1996. Mapping for germplasm collections: Site selection and attribution. *In* Proc. 3rd Int. Conf. Integrating GIS and Environ. Model., Santa Fe, NM. 21–25 January. NCGIA, Santa Barbara, CA (Available on-line at http://www.ncgia.ucsb.edu/conf/SANTA FE CD-ROM/main.html).

Jones, P.G., S.E. Beebe, J. Tohme, and N.W. Galwey. 1997. The use of geographic information systems in biodiversity exploration and conservation. Biodivers. Conserv. 6:947–958.

Rumbaugh, M.D., and B.M. Pendery. 1990. Germination salt resistance of alfalfa (*Medicago sativa* L.) germplasm in relation to subspecies and centers of diversity. Plant Soils 124:47–51.

Vavilov, N.I. 1992. Problems concerning new crops. p. 256–285. *In* V.F. Dorofeyev (ed.). Origin and geography of cultivated plants. Collected papers by N.I. Vavilov. (In Russian.) Cambridge Univ. Press, Cambridge, UK.

Vavilov, N.I. 1992. The phyto-geographic basis for plant breeding: Studies of the original material used for plant breeding. p. 316–366. *In* V.F. Dorofeyev (ed.) Origin and geography of cultivated plants. Collected papers by N.I. Vavilov. (In Russian.) Cambridge Univ. Press, Cambridge, UK.

# 7

# Predicting Species Distributions Using Environmental Data: Case Studies Using *Stylosanthes* Sw.

**Mark C. Sawkins**

*The Royal Botanic Gardens*
*Kew, Richmond, United Kingdom*
*and University of Birmingham*
*Birmingham, United Kingdom*

**Nigel Maxted**

*University of Birmingham*
*Birmingham, United Kingdom*

**Peter G. Jones**

*CIAT*
*Cali, Colombia*

**Roger Smith**

*Kew Seed Bank*
*West Sussex, United Kingdom*

**Luigi Guarino**

*IPGRI Regional Office for the Americas*
*Cali, Colombia*

The Convention of Biodiversity (UNCED, 1992) sets governments the task of conserving native biodiversity. An inventory of what exists and where it can be found is a logical and proper first step in this process. Plant exploration is far from complete. The rate of species extinction and genetic erosion prevents a full inventory being made before decisions are taken on conservation measures. The ecogeographic approach coupled with the power of Geographic Information Systems (GIS) software offers the opportunity of integrating the little that is already known from preexisting technical information about occurrence of species, with other environmental aspects of the habitat for which greater detail is available, such as climate, soils, and vegetation types. Thus, seeking correlations between occurrence of species and environmental variables and so predicting other areas of likely occurrence is of value both to the planning of ex situ and in situ conservation and may provide a greater understanding of the spatial distribution of genetic diversity.

Copyright © 1999. ASA and CSSA, 677 S. Segoe Rd., Madison, WI 53711, USA. *Linking Genetic Resources and Geography: Emerging Strategies for Conserving and Using Crop Biodiversity.* CSSA Special Publication no. 27.

Geographic Information Systems software, considered to be a novel technique for use in plant genetic conservation, is gradually becoming more widely utilized, as in studies examining the distribution of crops, wild relatives of crops and wild species. They can be used for example, in assessing gaps in collection, identifying suitable areas where species may be introduced and identifying potential sites for the conservation of species that cannot be placed in long-term storage (ex situ) for one reason or another. Franklin (1995) reviews the various methodologies that have been used in the predictive mapping of vegetation. More specifically Booth et al. (1989), Booth (1990), Sutherst and Maywald (1985), Walker and Cocks (1991), Carpenter et al. (1993), Huntley et al. (1995), Sutherst et al. (1995), Guarino et al. (1999, see Chapter 1), and Afonin and Greene (1999, see Chapter 6) describe methodologies for the purpose of predicting the distributions of particular animal and plant species.

This case study has been undertaken to test the current utility of this approach for wild species conservation. *Stylosanthes* Sw. is a tropical herbaceous legume naturally distributed in the Americas, Africa and Southeast Asia (Williams et al., 1984; Edye, 1987). Currently 30 to 40 species are recognized. *Stylosanthes humilis* H.B.K., which occurs naturally in South and Central America, also is adventive to northern Australia. The primary center of diversity for this genus is South America, in particular Brazil. A secondary center of diversity exists in the Mexican-Caribbean area (Stace & Cameron, 1984; Edye, 1987). Lewis (1987) suggested that the Brazilian state of Bahía is the center of evolution and explosive radiation of a number of legume genera, *Stylosanthes* Sw. included. Species occur in a variety of habitats and conditions and can be classified as weeds. They are colonizers of disturbed and open ground. Most grow in areas that have a well-defined dry season, experiencing varying amounts of seasonal drought (Williams et al., 1984).

The genus *Stylosanthes* Sw. was chosen for the case study because:

1. Its introduction into agriculture is only recent. Selection has been the major breeding approach. Therefore it will be a good model for wild species.
2. The collapse of its accidental introduction in Australia led to a concerted collecting effort, by both the Australians and others, of not just the original species but many more. Therefore relatively large and contemporary data sets are available for use.
3. *Stylosanthes* Sw. as an important forage plant has been the subject of considerable taxonomic and ecogeographic study. A range of species in the genus has geographic distributions, which are both widespread and apparently distinct, but do overlap. Therefore there is a good background against which to evaluate any extrapolation from GIS. We can observe if the technique is sufficiently sensitive to discriminate relative small differences between species.

Four species within the genus will be used in this work: *S. guianensis* (Aubl.) Sw., *S. capitata* Vog., *S. viscosa* Sw. and *S. humilis* H.B.K. These were chosen first because large ecogeographic data sets exist for analysis. Second, these species have received great attention and therefore a large bibliography is available. Since the discovery of agronomic uses within the genus, certain species have become important in improving native pastures, both in South America and further afield, especially on more demanding soils (acid soils, soils with high Al content) in which other species find it difficult to grow successfully. Also, studies have shown that some species may be useful as a biological control of the cattle tick (*Boophilus microplus* Can). Certain species produce viscous secretions from hairs, which cover the plant. These secretions can kill larvae of the cattle tick within 24 h (Sutherst et al., 1982; 1988).

*Stylosanthes* Sw. has been used to great effect in improving pasture land in the subhumid and semiarid regions of northern Australia, Thailand, China, India, West Africa and in the South American countries of Colombia, Brazil, and Peru (Stace & Edye, 1984). In East and South Africa, species of *Stylosanthes* Sw. also have been evaluated for pasture improvement (Thomas & Sumberg, 1995). In Australia alone, an estimated 1 million hectares of pasture have been sown with commercial *Stylosanthes* cultivars (Chakraborty et al., 1996). From 1965 to 1985, 16 cultivars were commercially released worldwide (Edye, 1987). It seems likely that introduction to other homoclines will be equally economically beneficial.

This genus clearly represents a valuable resource to the farmer for improving pasture. This is reflected in the amount of work that has been undertaken on this plant group; research continues to the present day in the area of genetic resources. The species themselves exhibit wide environmental and morphological variation. Cultivars released to the farmer have been developed from the selection of particular ecotypes evaluated in field trials. With such a wide range of ecotypes, further collection may identify plants which could make a valuable contribution to the use of *Stylosanthes* Sw., particularly in marginal areas (soils low in nutrition and/or very acid) or in areas subject to severe infestation by anthracnose (*Colletotrichum gloeosporiodes*). Thus it would be particularly useful to have tools that help identify possible areas that are likely to contain plants possessing useful traits or adaptations.

## THE CONSERVATION AND USE OF *STYLOSANTHES* SW. GENETIC RESOURCES

The gene pool of the genus *Stylosanthes* Sw. has given rise to some important cultivars grown in many regions of the world. In Australia, for example, *S. humilis* H.B.K. was noted as a legume with potential for use as an animal feed as early as 1914 (Burt & Williams, 1975). The legume became naturalized in the northern part of the state of Queensland. Seed of this species is thought to have been transported in ship refuse from Brazil to the port of Townsville at the turn of the century (Crowder & Chheda, 1982). By the early 1970s Townsville stylo, as it is commonly known, covered an estimated 2 million hectares of grazing land (Chakraborty et al., 1996).

Two events caused a rapid decline in the use and production of seed of Townsville stylo in Australia. A prolonged wet season from 1973 to 1974 provided the perfect conditions for the spread of the fungal disease anthracnose (*Colletotrichum gloeosporioides*) and this decimated pastures containing Townsville stylo. Anthracnose is thought to have been accidentally introduced into Australia on infected *Stylosanthes* Sw. seed. It is one of the most devastating diseases affecting *Stylosanthes* Sw. sp. throughout the Americas where it is indigenous and in areas where species and cultivars have been introduced (Lenné & Calderon, 1984). At the same time, a dramatic crash in beef prices occurred in Australia that reduced farmers' profits and thus their ability to purchase seed (Hopkinson & Walker, 1984). These events led to the realization that the Australian beef industry could not rely solely on Townsville stylo. A concerted effort was made to discover "new" species of pasture legumes that were more resistant to anthracnose and tolerant of a wide variety of environments.

An Australian scientist, W. Hartley, made the first collections of *Stylosanthes* Sw. from 1947 to 1948 in South America (Argentina and Paraguay). Since the 1960s germplasm collecting by the Commonwealth Scientific and Industrial Research Organisation (CSIRO) Davies Laboratory began in earnest with further trips to South and Central America and the collection of naturalized populations of *S. humilis* H.B.K. in Australia. Other international institutes also began to collect germplasm of a number of species and from a variety of environments. Examples of such institutes are the Centro Internacional de Agricultura Tropical (CIAT, Colombia), the University of Florida, the Fondo Nacional de Investigaciones Agropecuarias (FONAIAP, Venezuela) and the Centro Nacional de Recursos Genéticos, (CENARGEN, Brazil). Major collectors of germplasm include H.S. McKee (Central America), D. Norris (West Indies, Brazil), R.J. Williams (Brazil, Bolivia, Argentina), A.E. Kretschmer (Costa Rica), L.A. Edye (Brazil), R.L. Burt (Brazil, Belize, Panama, Venezuela), R. Schultze-Kraft (Brazil, Colombia, Venezuela), R. Reid (Cuba, Argentina, Mexico, Colombia), L. Coradin (Brazil), D. Cameron (Brazil), I.B. Staples (South Africa, Zimbabwe, Tanzania, India) and P.J. Skerman (Sudan).

A number of international institutes conserve *Stylosanthes* Sw. germplasm. CIAT holds by far the most accessions, with 3607 accessions of *Stylosanthes* Sw. in trust, of which about 1400 are *S. guianensis* (Aubl.) Sw. The International Livestock Research Institute (ILRI), CENARGEN, CSIRO, the University of Florida and the U.S. National Plant Germplasm System (US-NPGS) hold other significant collections (SGRP, 1996), although in some instances this may reflect shared germplasm between institutes.

## NATURAL DISTRIBUTION OF *STYLOSANTHES* SW.
## USED IN THIS STUDY

The perennial species *S. guianensis* (Aubl.) Sw. has a wide distribution throughout South and Central America, between the latitudes 23°N lat and 27°S long (Williams et al., 1984). It is absent from the Amazon basin, northern Mato Grosso in Brazil and northern Bolivia and has not been reported from the Caribbean Islands, and the Yucatan in Mexico. It has a distribution that is sympatric to that of two other species, *S. viscosa* Sw. and *S. humilis* H.B.K., although the latter species are more restricted to the coastal areas of Venezuela and Brazil (Edye, 1987). *Stylosanthes viscosa* Sw. is a perennial species that also has a wide distribution. Although it is sympatric with *S. guianensis* (Aubl.) Sw. in most of its range, Williams et al. (1984) noted that some areas of its distribution in the Northern Hemisphere do not overlap. They are: Sonora and the tip of Baja California in Mexico, a restricted area in Texas, and a fourth area that includes the islands of Cuba, Dominica, and Jamaica.

The annual species, S. *humilis* H.B.K. has a wide distribution throughout Central and South America. It spreads far into the interior of Brazil and is considered a common species, found in Venezuela, Cuba, Mexico and Central America. *Stylosanthes capitata* Vog. is a perennial tetraploid species, the distribution of which, although wide, is rather broken. It only occurs in eastern Brazil and Venezuela, and is absent from the western side of South America. In eastern Brazil it is much more widespread, situated mainly in the northeast and southeast of Brazil (Sousa Costa & Schultze-Kraft, 1993). Both authors opine that in the last 10 yr the construction of roads and subsequent increased traffic of people and animals have allowed this species to spread into new areas in both countries thus modifying its natural distribution.

## ESTIMATING THE POTENTIAL DISTRIBUTION OF TAXA

The method used in this study is one that has been developed at CIAT. Jones et al. (1997) describe in full detail the methodology used here. Figure 7–1 illustrates the various steps involved.

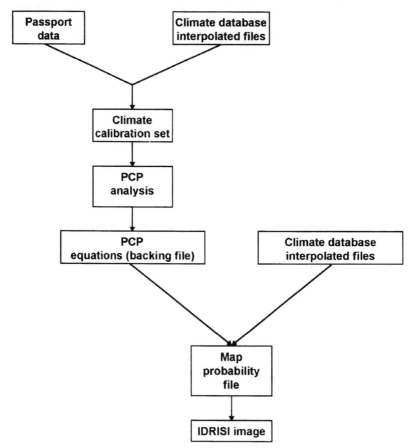

Fig. 7–1. Data flow for estimating potential distribution of *Stylosanthes* species.

First, passport data of the four species were assembled from the databases held at CIAT and CSIRO, with some data obtained from the online database held at CENARGEN, Brazil. Each location where *Stylosanthes* Sw. has been collected had an associated latitude, longitude and altitude. A file was created for each species consisting of the accession number, longitude, latitude and elevation of a location where an accession had been collected. In cases where a collector has not provided an elevation, it was necessary to consult gazetteers and/or maps to obtain this.

Next, the climate for each location was estimated. This resulted in the generation of a file that contains 12 mean monthly values of rainfall, temperature and diurnal temperature range (36 variables in total) for each location where *Stylosanthes* Sw. had been collected. These values were interpolated, using an inverse-square weighted distance, from data obtained at five of the closest meteorological stations. In addition, the temperature of each location was adjusted to take into account the elevation of the point of collection using a method developed by Jones et al. (1991).

The next step in the process was to take into account the differences in the timing of the seasons at different locations. The climate variables were transformed using a method developed by Jones et al. (1997). The seasons were aligned for all locations so that climates at different locations were not treated as dissimilar merely because their rainfall maxima and associated values of the other variables occur at different times of the year. These data were then further transformed on the rainfall values to ensure that the distribution of values was more normal. The square root transformation was used where $x \rightarrow \sqrt{x}$.

The climate calibration set was then put through a Principal Components Analysis (PCP) using the software package GENSTAT (Payne et al., 1987, p. 496–498) to identify fewer variables that account for most of the variation observed in the data. A multivariate-normal distribution was fitted to these principal components to enable the probability density corresponding to any climate to be calculated. This probability density indicates the frequency with which *Stylosanthes* Sw. was collected from areas with a climate of this type.

The same interpolation method used for the collection data was then used to estimate the climate associated at each location in a continent-wide data set based on the 10-min (18-km) grid from the National Oceanographic and Atmospheric Agency (NOAA) digital elevation model (NOAA, 1984). The climate for each grid point was aligned as before and expressed in terms of the principal components so that the associated probability density could be calculated. The results from this process were then plotted using a Fortran program to create an IDRISI image (Eastman, 1995) with a pixel corresponding to each grid point showing the probability distribution of the species in question throughout the continent or for a particular area.

## AN EXAMPLE OF ESTIMATING DISTRIBUTION FOR *STYLOSANTHES CAPITATA* VOG.

The procedures involved in estimating the distribution of a species are described using as an example *S. capitata* Vog. A file containing data on 311 accessions was collated, each accession having a latitude, longitude and elevation. These

Table 7–1. Climate variables for areas in which *Stylosanthes capitata* Vog. has been collected (data taken from the climate calibration file).

| Month | Rainfall (mm) | | | | | Temperature (°C) | | | | Diurnal temperature range (°C) | | | |
|---|---|---|---|---|---|---|---|---|---|---|---|---|---|
| | Minimum | Mean − standard deviation | Mean | Mean + standard deviation | Maximum | Minimum | Mean | Maximum | Standard deviation | Minimum | Mean | Maximum | Standard deviation |
| 1 | 77.0 | 160.8 | 224.4 | 287.9 | 453.0 | 19.6 | 23.7 | 28.7 | 1.8 | 5.6 | 10.1 | 15.3 | 1.8 |
| 2 | 90.0 | 139.4 | 199.5 | 259.6 | 383.0 | 20.0 | 23.7 | 28.9 | 1.8 | 5.8 | 10.2 | 14.7 | 1.7 |
| 3 | 73.0 | 115.2 | 172.6 | 230.1 | 286.0 | 19.8 | 23.7 | 28.2 | 1.9 | 5.8 | 10.3 | 14.5 | 1.6 |
| 4 | 19.0 | 63.7 | 96.5 | 129.2 | 170.0 | 19.1 | 23.5 | 28.1 | 2.0 | 6.2 | 10.9 | 15.4 | 1.7 |
| 5 | 9.0 | 16.8 | 38.9 | 61.1 | 152.0 | 17.1 | 22.8 | 27.5 | 2.4 | 6.4 | 12.2 | 17.5 | 2.3 |
| 6 | 1.0 | −5.6 | 20.2 | 46.1 | 103.0 | 15.4 | 22.1 | 27.0 | 2.9 | 6.5 | 13.5 | 20.3 | 3.0 |
| 7 | 1.0 | −8.2 | 17.9 | 44.0 | 123.0 | 14.0 | 22.1 | 28.3 | 3.5 | 6.8 | 14.1 | 21.2 | 3.2 |
| 8 | 1.0 | −1.8 | 18.0 | 37.8 | 105.0 | 13.9 | 23.1 | 29.3 | 3.5 | 6.7 | 14.3 | 21.5 | 3.1 |
| 9 | 3.0 | 12.2 | 34.5 | 56.8 | 101.0 | 16.2 | 24.4 | 29.3 | 2.8 | 6.7 | 13.5 | 20.4 | 2.6 |
| 10 | 35.0 | 56.3 | 91.1 | 125.9 | 167.0 | 18.1 | 24.7 | 29.1 | 2.3 | 6.8 | 12.3 | 17.9 | 2.1 |
| 11 | 71.0 | 127.1 | 165.9 | 204.8 | 335.0 | 19.0 | 24.3 | 28.7 | 2.0 | 6.5 | 10.9 | 16.0 | 1.7 |
| 12 | 72.0 | 165.2 | 231.6 | 297.9 | 453.0 | 18.9 | 23.8 | 28.5 | 1.9 | 6.0 | 10.1 | 15.1 | 1.7 |

data were then used to form the climate calibration set. Table 7–1 provides a description of the climate variables at the locations where *S. capitata* Vog. has been
collected. These climate data have been averaged after rotation to a standardized
date; therefore no information on the absolute dates of climatic events is retained.
Note that the rotation has placed the wettest period at the start of the newly defined
year. *Stylosanthes capitata* Vog. is known to be largely found in locations that exhibit a marked dry season lasting from 16 to 24 wk (4–6 mo) and extends into areas
that are semiarid in nature. This also is reflected in the types of habitat in which
plants have been found. *Stylosanthes capitata* Vog. has been collected from savanna
and dry tropical forest (Sousa Costa & Schultze-Kraft, 1993). Using 60 mm as the
rainfall below which a month is dry (Koppen, 1918), the data in Table 7–1 show
that the extreme low cases do, in fact, have about six dry months. On average, there
are 20 dry weeks (5 dry mo), but in wetter areas the dry season is not so clearly defined. Usually temperatures are constant throughout the year with an average temperature of about 23°C and a maximum of about 28°C. In the cooler areas, however, a marked seasonality can be seen in the temperature data, with temperatures
dropping during the dry season. The drier months throughout show a greater diurnal temperatures range. This is to be expected because of clear skies.

A PCP was then performed on the calibration set. Figure 7–2 plots the latent
vector values for the first four principal components of climatic variables against
the standardized months for *S. capitata* Vog. The first principal component accounts
for 40% of the variation encountered, is positively correlated with rainfall in the
6th and 7th mo, and negatively correlated with diurnal temperature difference in
the 8th mo. This component appears to reflect variation in the intensity of the dry
season, associated with opposite effects of temperature and diurnal temperature
range.

Principal Component 2, which accounts for 31% of the variation, is positively
correlated with rainfall in the 12th mo and negatively correlated with temperature
in the 8th mo. The third principal component, accounting for 15% of the variation,
is positively related to the dry season as is Component 1 but this time, with equiv-
alent effects of temperature and diurnal temperature range. The first and third
components are therefore well placed to differentiate climate types in the typical
areas. The fourth and fifth component each account for about 4% of the observed
variance. The chi-square for the two roots ($\chi^2 = 9.96$, $P > 0.01$) shows that they are
not equal. However, the first four components taken together account for about 90%
of the observed variance. Jones et al. (1997) found that using the first four components produced a clear probability mapping of *Phaseolus vulgaris* L. The analysis presented here was therefore restricted to the first four components.

## MAP OF POTENTIAL DISTRIBUTION FOR
### *STYLOSANTHES CAPITATA* VOG.

Figure 7–3 shows the results of the methodology employed to estimate the
potential distribution of *Stylosanthes capitata* Vog. Areas of high probability are
light to medium gray, areas of low probability are dark gray, and areas of no climatic similarity are colored black. It can be seen that many high climate probabil-

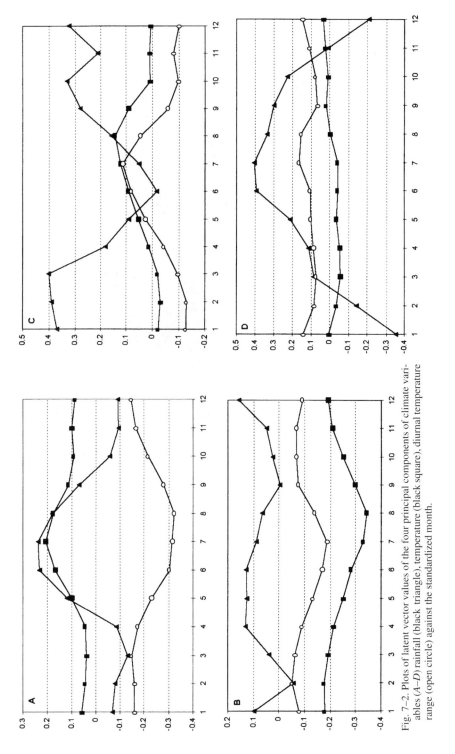

Fig. 7–2. Plots of latent vector values of the four principal components of climate variables (*A–D*) rainfall (black triangle), temperature (black square), diurnal temperature range (open circle) against the standardized month.

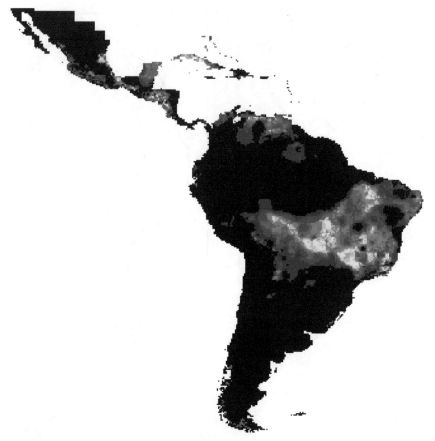

Fig. 7–3. Probability distribution for *Stylosanthes capitata* Vog. Areas of high probability are light to medium gray, areas of low probability are dark gray, and areas of no climatic similarity are colored black.

ity areas occur in southern Brazil. When the collection localities are overlaid on the probability image, some high probability areas are found where no collections have been made. Two of the most distinct are: northwestern Venezuela and the southern states of Brazil, especially near the border between Bolivia and Brazil (Mato Grosso and Mato Grosso do Sul).

## DISCUSSION

The probability maps generated for each species in this work match quite closely the natural distributions as described by Williams et al. (1984). Comparing the maps produced for the four species reveals that although in general similar areas are identified for all species, differences are observable. The potential distribution map for the example species, *S. capitata* Vog., extrapolates into areas where it has

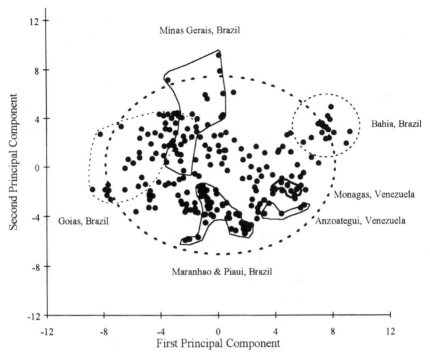

Fig. 7–4. Distribution of locations from which *S. capitata* Vog. Has been collected. Plot of principal components and 1 and 2.

so far not been found. It is known to occur only in Brazil and Venezuela. However, the map produced shows areas of high probability in Mexico, Honduras, Nicaragua, Cuba, Colombia, Bolivia and Argentina. This illustrates the fact that species may not always be found in areas possessing a suitable climate for growth. Other as yet unknown factors are influential in determining the geographical range of a species.

Another factor to consider is that many species are adapted to a variety of climates and environments. The procedure described above makes the assumption that *S. capitata* Vog. occurs in a continuous range of climates, which can readily be described by a multivariate normal distribution. Figure 7–4 shows a plot of the first two principal components for *S. capitata* Vog. The location points are spread throughout the two standard deviation ellipses, but points fall outside of this ellipse with perhaps some clumping, which may be evidence for the existence of different populations or ecotypes. Thomas and Grof (1986) describe four ecotypes that can be distinguished in this species. The first is a type that is found in central Brazil (e.g., Minas Gerais), which is early flowering, has high seed production and low dry matter yield. Another group occurs in Maranhao and Piauí, and is an intermediate type with lower seed production but larger dry matter yields, while a third type from Bahía produces high yields of dry matter as the growing season is longer. Finally, a fourth type occurs in Venezuela, is similar to that found in Bahía, but flowers earlier. From the plot of the first two principal components in Fig. 7–4, both the outliers have been highlighted as have the accessions collected from the areas which

are related to particular morphological types. From the climate data we may be see-
ing some evidence of clumping, although accessions collected from Bahía in Brazil
are spread throughout the area of the two standard deviations ellipse.

To investigate whether the species under study are composed of composite
populations, the data for each species has been hierarchically clustered into locations of similar climates, using the group average method (Payne et al., 1987, p.
496–498). Each climate cluster can then be mapped (Jones et al., 1997) separately
to identify areas of similar climates. Including other environmental variables may
refine this analysis. Soil characteristics may be important for modifying the distribution, as may habitat disturbance because *Stylosanthes* Sw. are pioneer species.


## CONCLUSION

Other workers are examining the mapping of the potential distributions of
plants and animals for a variety of reasons. This chapter sets out the results of a study
dealing with one plant genus, attempting to further understanding of the patterns
of distribution for the various species chosen. For many wild species little information (apart from a collection locality) exists concerning specific environmental
requirements influencing distribution. This is particularly true of data held in
herbaria, where in extreme cases perhaps only a few examples of a species may be
held. The method described in this chapter can be quickly applied to different
species. For our purposes, this method is useful for identifying new areas to focus
or plan collecting missions or for in situ conservation. Combining the climate
mapping technique with genetic information also may provide a clearer understanding of the spatial distribution of genetic diversity in *Stylosanthes* Sw. in the
Americas.

This method has been validated with *P. vulgaris* L. where weedy types have
been found in areas with climates corresponding to a high probability density and
areas have been identified where, although no collections have been made, presence of the species has been reported (Jones et al., 1997).


## REFERENCES

Booth, T.H. 1990. Mapping regions climatically suitable for particular tree species at the global scale.
    Forest Ecol. Manage. 36:47–60.
Booth, T.H., J.A. Stein, H.A. Nix, and M.F. Hutchinson. 1989. Mapping regions climatically suitable
    for particular species: an example using Africa. Forest Ecol. Manage. 28:19–31.
Burt, R.L., and W.T. Williams. 1975. Plant introduction and the *Stylosanthes* story. Australian Meat Res.
    Comm. 25:1–23.
Caldecott, J.O., M.D. Jenkins, T.H. Johnson, and B. Groombridge. 1996. Priorities for conserving global
    species richness and endemism. Biodivers. Conserv. 5:699–727.
Carpenter, G., A.N. Gillison, and J. Winter. 1993. DOMAIN: A flexible modelling procedure for map-
    ping potential distributions of plants and animals. Biodivers. Conserv. 2:667–680.
Chakraborty, S., D.F. Cameron, and J.A. Lupton. 1996. Management through improved understanding—
    a case history of *Stylosanthes* anthracnose in Australia. p. 603–619. *In* S. Chakraborty et al. (ed.)
    Pasture and forage crop pathology. ASA, Madison, WI.
Crowder, L.V., and H.R. Chheda. 1982. Tropical grassland husbandry. Longman, London.
Eastman, J.R. 1995. IDRISI for Windows. A users guide, version 1.0. Clark Univ.

Edye, L.A. 1987. Potential of *Stylosanthes* for improving tropical grasslands. Outlook Agric. 16:124–130.

Franklin, J. 1995. Predictive vegetation mapping: geographic modelling of biospatial patterns in relation to environmental gradients. Progr. Phys. Geogr. 19:474–499.

Guarino, L., N. Maxted, and M. Sawkins. 1999. Analysis of georeferenced data and the conservation and use of plant genetic resources. p. 1–24. *In* S.L. Greene and L. Guarino (ed.) Linking genetic resources and geography: Emerging strategies for conserving and using crop biodiversity. CSSA Spec. Publ. 27. ASA and CSSA, Madison, WI.

Hopkinson, J.M., and B. Walker. 1984. Seed production of *Stylosanthes* cultivars in Australia. p. 433–449. *In* H.M. Stace and L.A. Edye (eds.) The biology and agronomy of *Stylosanthes*. Acad. Press, Sydney, Australia.

Huntley, B., P.M. Berry, W. Cramer, and A.P. McDonald. 1995. Modelling present and potential future ranges of some European higher plants using climate response surfaces. J. Biogeog. 22:967–1001.

Jones, P.G., D.M. Robison, and S.E. Carter. 1991. A GIS approach to identifying research problems and opportunities in natural resource management. p. 73–110 *In* CIAT in the 1990s and beyond, a strategic plan. Supplement. CIAT, Cali, Colombia.

Jones, P.G., N.W. Galwey, S.E. Beebe, and J. Tohme. 1997. The use of geographical information systems in biodiversity exploration and conservation. Biodivers. Conserv. 6:947–958.

Koppen, W. 1918. Klassifikation der Klimate nach Temperatur, Niederschlagund Jahreslauf. Petermanns Geographische Mitteilungen 64:193–203, 243–248.

Lenné, J.M., and M.A. Calderon. 1984. Disease and pest problems of *Stylosanthes*. p. 279–294. *In* H.M. Stace and L.A. Edye (ed.) The biology and agronomy of *Stylosanthes*. Acad. Press, Sydney, Australia.

Lewis, G. 1987. Legumes of Bahia. R. Bot. Gardens, Kew, New Zealand.

National Oceanographic and Atmospheric Administration. 1984. TGP-006 D. computer compatible tape. NOAA, Boulder, CO.

Payne, R.W., P.W. Lane, A.E. Ainsley, K.E. Bicknell, K.E. Digby, P.G.N. Harding, S.A. Leech, H.R. Simpson, A.D. Todd, P.J. Verrier, R.P. White, J.C. Gower, G. Tunnicliffe Wilson, and L.J. Paterson. 1987. Genstat 5 reference manual. Clarendon Press, Oxford, UK.

Sousa Costa, N.M.S., and R. Schultze-Kraft. 1993. Biogeographia de *Stylosanthes capitata* Vog. y de *Stylosanthes guianensis* Sw. var. *pauciflora*. Pasturas Trop. 15:10–15.

System-Wide Genetic Resources Programme. 1996. Report of the internally commissioned external review of the CGIAR genebank operations. Int. Plant Genet. Resour. Inst., Rome, Italy.

Stace, H.M., and D. F. Cameron. 1984. Cytogenetics and the evolution of *Stylosanthes* Sw. p. 49–69. *In* H.M. Stace and L.A. Edye (ed.) The biology and agronomy of *Stylosanthes*. Acad. Press, Sydney.

Stace, H.M., and L.A. Edye. 1984. The biology and agronomy of *Stylosanthes*. Acad. Press, Sydney.

Sutherst, R.W., R.J. Jones, and H.J. Schnitzerling. 1982. Tropical legumes of the genus *Stylosanthes* immobilise and kill cattle ticks. Nature (London) 295:320–321.

Sutherst, R.W., and G.F. Maywald. 1985. A computerised system for matching climates in ecology. Agric. Ecosyst. Environ. 13:281–299.

Sutherst, R.W., L.J. Wilson, R. Reid, and J.D. Kerr. 1988. A survey of the ability of tropical legumes in the genus *Stylosanthes* to trap larvae of the cattle tick, *Boophilus microplus*. Austral. J. Exp. Agric. 28:473–479.

Sutherst R.W., G.F. Maywald, and D.B. Skarratt. 1995. Predicting insect distributions in a changed climate. p. 59–91. *In* R. Harrington and N.E. Stork (ed.) Insects in a changing environment. Acad. Press, London.

Thomas, D., and B. Grof. 1986. Some pasture species for the tropical savannas of South America. I. Species of *Stylosanthes*. Herb. Abstr. 56:445–454.

Thomas, D., and J.E. Sumberg. 1995. A review of the evaluation and use of tropical forage legumes in sub-Saharan Africa. Agric. Ecosyst. Environ. 54:151–163.

U.N. Conference on Environment and Development. 1996. Convention on biological diversity. UNCED, Geneva, Switzerland.

Walker, P.A., and K.D. Cocks. 1991. HABITAT: A procedure for modelling a disjoint environmental envelope for a plant or animal species. Global Ecol. Biogeog. Lett. 1:108–118.

Williams, R.J., R. Reid, R. Schultze-Kraft, N.M. Sousa Costa, and B.D. Thomas. 1984. Natural distribution of *Stylosanthes*. p. 73–101. *In* H.M. Stace, and L.A. Edye (ed.) The biology and agronomy of *Stylosanthes*. Acad. Press, Sydney.

# 8 Institutional Adoption of Spatial Analytical Procedures: Where Is the Bottleneck?

**John D. Corbett and Paul Dyke**

*Texas A & M University*
*Temple, Texas*

Geographic Information Systems (GIS) technology has tremendous potential to support research in agricultural and natural resource management. However, adoption of the technology requires much more than the purchase of hardware and software, and the hiring of technicians. Spatial analytical procedures and the databases that support them often challenge traditional and strongly institutionalized processes. "Adopting" spatially referenced information requires intellectual acceptance of a new organization of the quantifying methods and data and a new paradigm in the logical processes of analysis. The range of institutional response is dramatic, from uncertainty and no direct response, to altering project goals, to progressing beyond acceptance and demanding more results. Simply stated, institutional adoption of spatial analytical procedures depends on the clear demonstration of methods, data, and opportunities. The bottleneck lies not with the data or spatial technology but rather with an understanding of the new paths of analyses available to link spatially explicit analysis to institutional assessment of alternatives and the implementation of decisions.

Geographic Information Systems has become a catchall term, signifying state-of-the-art spatial analysis as well as a money and effort sink with expectations of payback in "10 years." The full spectrum of adoption experiences reflects the nature of any "new" technology but all appear to have a common thread: success or failure, explanations point to the gap between known, explicit analytical capability, and the institutional adoption and use of the results.

An analogy might be that of a dam, holding back ever-increasing amounts of water (spatial data and software/hardware capability), with many tested and untested turbines (demand), but also with a constricted or underutilized intake (spatial analytical procedures) capacity. Removing this constriction to flow—the bottleneck—just as in the design and use of intakes on a dam—requires careful planning with stages (possibly iterative) for design, testing, and implementation. The turbines, in this analogy, represent a full spectrum of a dynamic demand, literally giving feedback both for timing and volume, to the analytical capacity as represented by the intakes.

Geographic literacy, here defined as the awareness of the spatial context and the opportunities for spatial analysis, is the critical demand component that determines the rate of transfer of spatial analytical capability to the institutional adoption and use of the results. Craig and Johnson (1997) provide three terms that enable a debate on the institutional success of GIS: *efficiency, effectiveness, and equity*. Efficiency is doing standard things quicker and cheaper. Effectiveness is doing the old things better—improving quality—and doing new things to improve decision-making. Equity is being fairer in dealing with people and organizations. About the "GIS" community, Obermeyer (1994) writes "We frequently speak among ourselves of the value and benefits of geographical information systems, firm in the belief that GIS can help improve decision-making in both the public and private sectors." It is apparent that within the GIS community, equity in decision-making is a key motivational force driving efforts to institutionalize GIS.

The authors are of the opinion that the "package" nature or "object" nature of spatial data is one of the least understood by decision-makers. If the objective of a spatial analysis included equity in decision-making, then we can assume both efficiency and effectiveness in the spatial analysis and more important, geographically literate decision-makers. Equity issues require a fundamental understanding of the potential to bundle together heterogeneous information into objects. Once bundled, these objects become the basis for resource allocation decisions that account for previously impossible combinations of information. For example, if the policy objective is clean rivers, then creating objects that distinguish firm size (for capital access), soil erodibility, landscape position, and agroclimatic environment (including land use), insure an evaluation with the ability to address equity issues in the population.

The object structure permeates all three of the terms listed. Understanding this new paradigm requires the recognition that data accessed in spatial context is "flipped" in its logical structure. Historically, policy and decision methodologies are structured by subject matter and links are made between subjects (themes) to provide integrated analyses. *In the spatial paradigm, information is processed not by subject but by object.* Information is assembled and displayed by packaging many themes together to create an "object" [i.e., everything we know about a point (or area) in space]. This analytical methodology is consistent with much of the new software and hardware technologies surfacing in the information age such as object-oriented programming and object-oriented database. Synthesis occurs when these objects are linked to each other in geographic space. This means analytical descriptions are based on representative or typical objects aggregated into meaningful clusters rather than on statistical averages and deviations of disconnected thematic populations. Let us look more closely into how this spatial paradigm "plays out" in analyzing the institutional success of GIS.

## UNDERSTANDING THE POTENTIAL OF SPATIAL ANALYTICAL PROCEDURES

Tosta (1996) comments that "…trying to put a new technology in an old organization and successfully use it is damn hard." Institutional use of GIS to automate standard activities in the hope of making things quicker and cheaper (efficiency) is often an example of negative experiences. Institutions seeking efficiency are typically urban or county survey offices. Digitizing and verifying maps is expensive and time consuming. Institutional adoption of GIS for efficiency may need 10 yr for payback. Geographic Information Systems can certainly improve efficiency and, for some operations, efficiency, even with a 10-yr horizon, is sufficient. At first glance efficiency translates into being able to do more in less time and at less cost—a process understood by all. However, one can easily overlook the new analytical options associated with the time elements. With the increase in efficiency, the decision-maker can ask questions of the spatial decision support system not previously considered because the complexity of the analysis was considered beyond the analytical capacity in the decision time window. The bottleneck, even at the level of improving efficiency, is training the decision-maker to recognize the spatial analytical possibilities and refine the questions to exploit the spatial nature of the question.

Effectiveness is improved by reducing the number of steps between the question and the answer. Just as a "TV dinner" (an object) can shorten the process of feeding the body, preprocessing of data into spatial objects can increase the effectiveness of decision-making. The object nature of spatial analytical capability means that the grouping of objects can be specifically related to the question. In other words, classifications can be made to reflect discriminatory characteristics specific to the question (Corbett, 1996, unpublished data). Previously defined classifications are no longer needed early in the method (typical in the traditional, analog approach). A priori classifications are less useful because objective specific groupings can be made. Getting decision-makers (and database makers) to delay object aggregation (the classification) in the methodological design can be a major bottleneck in proper implementation of spatial analytical procedures. A delay in the aggregation of objects demonstrates an understanding of the uniqueness of spatial analytical procedures. Examples of appropriate preprocessing and delayed clustering are given below.

Agricultural and environmental uses of GIS are typically good at exploiting the character of effectiveness. In agricultural research, a GIS is capable of "doing old things better" and improving the quality of the results. A spatial database of climate surfaces (e.g., monthly macroscale spatial descriptions of long-term climatic normals created from interpolated meteorological station records) can be used to quickly and effectively portray adaptation zones [e.g., maize (*Zea mays* L.) variety germplasm adaptation] with far better quality than traditional paper map or analog efforts. These first results are analogous to the agroecological adaptation zones traditionally used but describe more precisely the geographic adaptation of the germplasm in question. In the following example, we selected a site in Kenya and queried our databases for areas in Africa with similar growing season characteristics (Fig. 8–1).

The example shown in Fig. 8–1 is provided to demonstrate the construction of a simple empirical model: a point and the area similar to it. If Kianjuki was a field trial location, it might be of interest to quickly know, within selected constraints, what areas of Africa might be similar to Kianjuki. The sense of being representative of a larger area is important not only to assess appropriate areas for further evaluation and possibly even a spatial sampling strategy, but also to prioritize research efforts. In any case, the ability to rapidly (with emphasis on the rapid characteristic) explore geographic expressions of similarity in environment is thought to be crucial to the expansion of the demand for spatial analytical procedures.

Doing old things better does not guarantee to open necessary and available "intake valves" required to institutionalize the use of spatial analytical techniques in agricultural and environmental research. Geographic literacy (defined as the awareness of the spatial context and the opportunities for spatial analysis) is the critical component in the transfer of known spatial analytical capability and the subsequent institutional adoption and use of the results. There may be a substantial training component necessary to build up this demand. Training targeted at senior administrators needs to reflect the results of an institution-specific analysis of practical information needs.

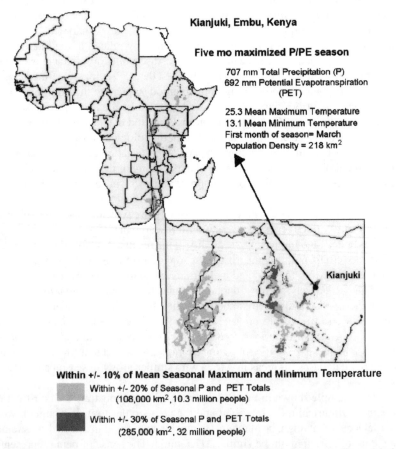

Fig. 8–1. The zone in Africa with similar growing season characteristics, for one specific model of the growing season, and the number of people living in the resultant zone.

If, as Craig and Johnson (1997) assert with respect to the use of GIS that, "The bigger payoff is in doing things that had been impossible or impractical before," agricultural and environmental challenges should find no shortage of examples.

## INSTITUTIONAL ADOPTION: EXAMPLES WITH COMMENTARY ON DEVELOPING DEMAND

### An Example from the World Health Organization

The World Health Organization (WHO) is combating "highland malaria" in East Africa. Highland malaria is defined as epidemic malaria outbreaks in non-malaria endemic areas. Because malaria is not endemic, the human inhabitants do not build up immunities and the population is described as "immunologically naïve." With population growth, land-use change (deforestation amongst others) and the specters of climate change on the horizon, normal intermittent warm years and subsequent malaria outbreaks cause a significant toll. When morbidity and mortality affect the productive age cohorts (15–50 yr), significant human and economic damage accrues.

In May 1996, epidemiologically trained scientists gathered in Addis Ababa to explore a regional initiative to address highland malaria. Epidemiologists are— in many ways— trained to think spatially and they routinely include georeferencing in data collection strategies. A brief presentation using Fig. 8–2 altered the workshop agenda and subsequent research effort since the challenge faced by the highland malaria team had a significant spatial character that could be reorganized into a more spatial, analytical, framework. Being spatially inclined, epidemiologists readily accepted the methods behind this first-order characterization and have subsequently employed the same databases to make far more specific analyses of the target zone (e.g., Lindsay et al., 1998). This application was successful because the program design team members were already, in effect, seeking information to make their effort more efficient, more effective, and certainly more equitable. They already had "unsatisfied" demand and the presentation of the Spatial Characterization Tools (Corbett & O'Brien, 1997) precursor (The Data Exploration Tool, Corbett et al., 1996) provided an opportunity to rapidly exploit available spatial analytical procedures.

### Texas A&M IMPACT Group

"The Texas A&M University System Agriculture Program's Impact Assessment Group provides a multidimensional capability to assess the economic, environmental, and societal impacts of change. Using a unique georeferenced framework that links interactive models, databases, and analytic methods, the group evaluates the impact of technology, policies or regulations, and related factors such as demographics or markets" (Clarke, 1997).

With the above statement, written as part of a promotional packet, wholehearted adoption of a spatial structure to agriculture and environmental impact assessment is in evidence at Texas A&M University. A spatial context is fundamental to the group's approach, providing the linkages between models from many disciplines (both for inputs to models and as a place to archive output) and "institutionalizing" the results. Notice that GIS is not mentioned nor should it be—the tools, GIS, are not nearly as important as the concept. The concept was not easy to incorporate into the group's objectives. More than a year of regular discussions took place before the previous paragraph was written. Emphasis is placed on the vision behind the awareness of spatial heterogeneity and developing spatial analytical procedures to meet a demand. This ensures not only a level of cooperation between spatial scientists and others with less experience with that perspective, but the group's potential results, from the outset, are communicated to administrators with a fundamentally spatial character. This does not mean that all results require spatial analytical procedures, rather that spatial analytical procedures are fundamen-

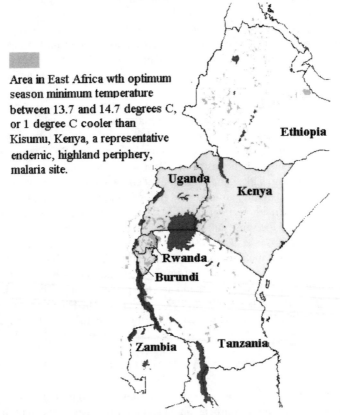

Area in East Africa with optimum season minimum temperature between 13.7 and 14.7 degrees C, or 1 degree C cooler than Kisumu, Kenya, a representative endemic, highland periphery, malaria site.

Fig. 8–2. Use of spatial analysis expanded the opportunities and knowledge base for the World Health Organization's effort to mitigate epidemic malaria in the Highlands of East Africa. Figure 8–2 was presented as a first approximation of the "target" zone for highland malaria activities (Corbett, 1996). From this analysis, new challenges to the highland malaria team were made more clear, their target zone for prevention and epidemic response activities is spatially highly disaggregated. The ramifications for monitoring malaria outbreaks and responding to them are severe as much of East Africa has a poor rural road system, a system that suffers additional problems during the rainy (and thus malaria season), as well as poor to nonexistent telephone systems. Optimum season is defined as the five consecutive months that maximize precipitation/potential evapotranspiration.

tally part of the package and will be brought to bear when efficiency, effectiveness, or equity demand them.

## ADDING IN SIMULATION MODELS—THE POWER OF WHAT IF?

Computer simulation models, especially sophisticated plot level models (e.g., EPIC or the CERES suite of crop models), are seldom seen as tools for the decision-maker whose needs are more aggregate. A plot scale model connected to spatial data can contribute to an analysis of the potential target area and subsequent impact of new germplasm. With this information, priorities can be better evaluated and research investments more accurately targeted.

Simulations allow us to alter the components of the objects (resource use or management) to evaluate possible scenarios reflecting a change in technology, the environment, policies, or other decision processes. The spatial environment then allows for "re-grouping" to pursue change in equity issues. Decisions taken against particular object groupings can be evaluated for their accuracy as well as impact. The construct delivers the potential to evaluate impacts both ex-ante or ex-poste.

The Rockefeller Foundation, in an ongoing effort to improve the spatial analytical capability of decision-makers, supported an effort to link GIS and simulation modeling. The objective was the creation of a method and a database that described the relative performance of maize germplasm over the whole of the East African region (Fig. 8–3). The resultant digital database (Collis & Corbett, 1998, unpublished data), contains not only the mean yield from the 30 yr of simulation, but also all the additional information provided by the simulation model. For example, the amount of N required to achieve the simulated yield. Such information could be interpreted with respect to necessary N replacement back into the soil.

Simulation models allow for the rapid evaluation of "what if" questions relative to some environmental change events. Best management practices and most appropriate germplasm (from simulations), in anticipation of a drier than normal season, can be invaluable information to disaster mitigation as well as development assistance (Collis & Corbett, 1998, unpublished data). Moreover, with the development of more demand for this kind of spatial procedure linked with simulations, there will be tremendous value added to both the spatial databases and the simulations models.

## CONCLUSION

Not all experiences are as positive as the previous examples. Some institutions do not respond to the efficiency, effectiveness, or equity promised by spatial analytical procedures. As proposed, most result from failure to increase the demand at the level where institutional decisions are taken. To overcome geographic illiteracy, the presentation of results from the "GIS Laboratory" must be carefully constructed. In the author's experiences, our failures occur when we assume more geographic literacy than actually exists. Somers (1996) writes "For example, initial impressions may have been that GIS would benefit the entire organization, but the

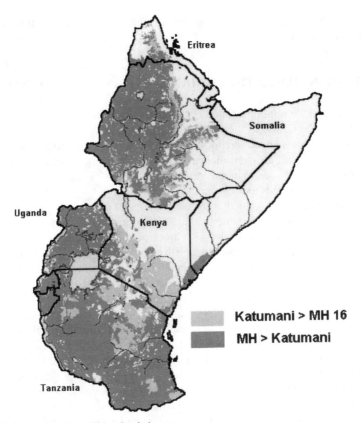

Fig. 8–3. Results from East Africa simulation

reality is that it can be implemented in limited places at the present time." In an effort to demonstrate the many potentials of spatial analytical procedures, it is possible to imply the unbelievable, however spatial analytics does not solve all problems. "It is important to be realistic about organizational constraints as well as to be able to recognize and act on opportunities when they occur" (Somers, 1996). The audience must be understood and addressed accordingly.

Presenting the results of the analysis in a sufficiently clear and concise manner and preparing the audience for the implications the results propagate may require substantial investment in "training" the target audience, and it will require an investment into the presentation of the results.

In the design of an agricultural or environmental GIS, there needs to be careful consideration in each step (planning, design, and implementation phase) of the level of geographic literacy in the target audience. It is crucial to evaluate the "geographical literacy" of the institution. The level of geographic literacy will determine the investment needed to present spatially explicit analytical results.

In some cases, results will require substantial retrofitting—literally educating the users about the information to improve their capability to request information. The difficulty of this step should not be underestimated. Creating a logical presentation that clearly introduces the potential of spatial analytical procedures is not the same presentation as describing methods and results. Aiming too high or assuming more geographic literacy than is present is a sure way to be caught wondering, years later, why the use of GIS is so minimal in an institution.

In other cases, the potential exists to create rather simple demonstrations of the power of spatial analytical procedures, and then have an audience of senior administrators become so enraptured with the potential that resources become overtaxed and results are demanded at ever increasing rates. Our point is simple. The spatial analytical procedures are sufficiently developed and the GIS software sufficiently robust that the bottleneck lies in the demand for these services. The acceptance of the result and knowing what to ask next is the key and this requires geographic literacy—geographic literacy being the ability not to be concerned with the "how" nearly as much as the ability to perceive improved equity on account of spatial analytical procedures.

One clear advantage agriculture and environmental efforts have over our "urban" oriented compatriots in GIS, is that typically a substantial amount of information is readily available from public sources. This information includes both databases and models. For example, digital elevation models at a 1-km resolution are freely available and cover the earth. A host of institutions with mandates to create spatial databases have been methodically building a suite of public domain data that, once assembled, can provide a small GIS project a substantial database free or nearly free of charge. For example, the Integrated Information Management Laboratory at the Blackland Research Center (Texas A&M Univ. System) has compiled a series of spatial databases (focused initially on Africa, but we have a growing capability for the Americas, including the USA and we are working to create a database for Asia) and query tools into a product we call the Spatial Characterization Tool (Corbett & O'Brien, 1997). A related product, the Country Almanacs (Corbett et al., 1998), supply similar agricultural and environmental information and do not require any additional software licenses.

Our approach to an integrated, spatial information system capitalizes on this investment in spatial data, simulation models, and GIS software. For example, the infrastructure data used in the "malaria" example could contribute to an analysis of market costs for agricultural products. These "value added" characteristics only accrue if combined in an accessible and flexible environment. Simulation models, historically isolated in the realm of the analytical scientist, promise to grow in value as a synthesizing tool. A spatial environment serves to integrate the spectrum of scales and disciplines (our objects) because the spatial environment provides the institution for relating inputs and outputs. This constructed "memory" serves to identify gaps in our data and understanding, which fosters research, contributes to the elimination of redundancies, targets resources more efficiently, and encourages exante and ex-post impact evaluation.

The level of geographic literacy in the decision-makers is the wild card. The responsibility to increase the demand for spatial analytical procedures lies primarily with those who understand its unique characteristics and potential. As stated, a key advantage to agricultural and environmental applications is in the ability to borrow heavily from readily available databases and models. This borrowing will take an institution to a certain point, beyond which there needs to be a careful focus on communicating both the specific nature of the results in terms of efficiency, effectiveness, and equity, and the larger contribution spatial analytical procedures can be made to the institutional mandate.

## REFERENCES

Clarke, N.P. 1997. Promotional brochure for The Texas A&M University System's IMPAC Group derived from impact methods to predict and assess change, annual report, September 30, 1997. Agric. Program, Texas A&M Univ. Syst., College Station, TX.

Corbett, J.D., R.F. O'Brien, E.I. Muchugu, and R.J. Kruska, 1996. The data exploration tool. A joint ICRAF, UNEP, ILRI, The Rockefeller Found., and Texas A&M Univ. Syst. CD-ROM Publ. (database and interactive interface tool). ICRAF, Nairobi, Kenya.

Corbett, J.D., and R.F. O'Brien. 1997. The spatial characterization tool—Africa. Vers. 1.0. Texas Agric. Exp. Stn. Blackland Res. Center Rep. no. 97-03.

Corbett, J.D. 1996. Some potential uses of GIS for malaria modeling in the East African Highlands. Presented to the World Health Organization Workshop on Highland Malaria, Addis Ababa, Ethiopia, May 1996.

Corbett, J.D. 1998. Classifying maize production zones in Kenya through multivariate cluster analysis. p. 15–26. In R. Hassan (ed.) Maize technology development and transfer: A GIS application for research planning in Kenya. CAB Int., CIMMYT, and the Kenya Agric. Res. Inst.

Corbett, J.D., S.N. Collis, R.F. O'Brien, B.R. Bush, E.I. Muchugu, R.A. Burton, R.E. Martinez, R. Gutierrez, and R.Q. Jeske. 1998. East African country almanacs. A resource base for characterizing the agricultural, natural, and human environments of Kenya, Ethiopia, Uganda, and Tanzania. A joint CIMMYT-Blackland Research Center CDROM Publ. Blackland Res. Center Rep. 98-08.

Craig, W.J., and D.D. Johnson, 1997. Maximizing GIS benefits to society. Geographic Information Systems. March, p. 14–18.

Lindsay, S.W., L. Parson, and C.J. Thomas. 1998. Mapping the ranges of relative abundance of the two principal African malaria vectors, Anopheles gambiae sensu stricto and An. Arabiensis, using climate data. Proc. R. Soc. London Biol. Sci. 265:847–854.

Obermeyer, N. 1994. GIS: A new profession? Prof. Geogr. 46(4):498–503.

Somers, R. 1996. How to Implement a GIS. Geographic Information Systems. January, p. 18–21.

Tosta, N. 1996. Organic organizations. Geographic Information Systems. March, p. 44–48.